AF886525

Linear Algebra and Differential Equations

BERKELEY MATHEMATICS
LECTURE NOTES

Volume 11

Linear Algebra and Differential Equations

Alexander Givental

American Mathematical Society
Berkeley Center for
Pure and Applied Mathematics

Editorial Board

Alexandre Chorin
L. Craig Evans
Alan Weinstein

2000 *Mathematics Subject Classification.* Primary 15–01;
Secondary 34–01, 35–01, 51–01.

Copying and reprinting. Individual readers of this publication, and nonprofit libraries acting for them, are permitted to make fair use of the material, such as to copy a chapter for use in teaching or research. Permission is granted to quote brief passages from this publication in reviews, provided the customary acknowledgment of the source is given.

Republication, systematic copying, or multiple reproduction of any material in this publication is permitted only under license from the American Mathematical Society. Requests for such permission should be addressed to the Assistant to the Publisher, American Mathematical Society, P. O. Box 6248, Providence, Rhode Island 02940-6248. Requests can also be made by e-mail to `reprint-permission@ams.org`.

© 2001 by Alexander Givental. All rights reserved.
The American Mathematical Society retains all rights
except those granted to the United States Government.
Printed in the United States of America.

∞ The paper used in this book is acid-free and falls within the guidelines
established to ensure permanence and durability.
Visit the AMS home page at URL: `http://www.ams.org/`

10 9 8 7 6 5 4 3 11 10 09

Contents

Foreword

To the student:

The present text consists of 130 pages of lecture notes, including numerous pictures and exercises, for a one-semester course in Linear Algebra and Differential Equations. The notes are reasonably self-contained. In particular, prior knowledge of Multivariable Calculus is not required. Calculators are of little use. Intelligent, hands-on reading is expected instead.

A typical page of the text contains several definitions. Wherever you see a word typeset in the font Sans Serif, it is a new term and the sentence is the definition.

A large portion of the text represents *Examples.* However, numerical illustrations or sample solutions to homework problems are rare among them. The examples are there because they are part of the theory, and familiarity with each one of them is crucial for understanding the material. Should you feel the need to see numbers instead of letters, you are welcome to substitute your favorite ones.

The notes are written in a concise, economical style, so do not be misled by the total size of the text: you can find there more material than you can think of. If you notice that reading a typical page takes less than an hour, it is a clear sign that your reading skills may need further polishing. Ask your instructor to give you some hints. Perhaps they will sound like this:

"Have you found out how the first sentence in a section implies the next one, the third one — follows from the second one, and so on?.. Have you checked that the statement of the theorem does not contradict the examples you keep in mind?.. Having done with this, try exercises ... Do not give up a problem before you are sure you know exact meaning of all technical terms it involves ... To make sure, write down their definitions in complete sentences ... "

If nothing helps, you are probably reading the wrong half of this Foreword.

To the instructor:

The lecture notes correspond to the course "Linear Algebra and Differential Equations" taught to sophomore students at UC Berkeley. We accept the currently acting syllabus as an outer constraint and borrow from the official textbooks two examples, [1] but otherwise we stay rather far from conventional routes.

In particular, at least half of the time (Chapters 1 and 2) is spent to present the entire agenda of linear algebra and its applications in the $2D$ environment; Gaussian elimination occupies a visible but supporting position (section 3.4); abstract vector

[1] "Competing species" from Boyce – DiPrima's *Elementary Differential Equations and Boundary Value Problems* and "Error-correcting codes" from *Elementary Linear Algebra with Applications* by R. Hill

spaces intervene only in the review section 3.7. Our eye is constantly kept on *why?*, and very few facts [2] are stated and discussed without proof.

The notes were conceived with somewhat greater esteem for the subject, the teacher and the student than is traditionally anticipated. We hope that mathematics, when it bears some content, can be appreciated and eventually understood. We wish the reader to find some evidence in favor of this conjecture.

[2] The fundamental theorem of algebra, the uniqueness and existence theorem for solutions of ordinary differential equations, the Fourier convergence theorem and the higher-dimensional Jordan normal form theorem.

CHAPTER 1

Geometry on the Plane

1.1. Vectors

Vectors is a mathematical abstraction for quantities, such as forces and velocities in physics, which are characterized by their magnitude and direction.

1.1.1. Definitions. A directed segment $\vec{AB}$ on the plane is specified by an ordered pair of points — the tail A and the head B. Two directed segments $\vec{AB}$ and $\vec{CD}$ are said to represent the same vector if they are obtained from one another by translation. In other words, the lines AB and CD must be parallel, the lengths $|AB|$ and $|CD|$ must be equal, and the segments must point out the same of the two possible directions.

A trip from A to B followed by a trip from B to C results in a trip from A to C. This observation motivates the following definition of the vector sum $\mathbf{w} = \mathbf{v} + \mathbf{u}$ of two vectors $\mathbf{v}$ and $\mathbf{u}$: if $\vec{AB}$ represents $\mathbf{v}$ and $\vec{BC}$ represents $\mathbf{u}$ then $\vec{AC}$ represents their sum $\mathbf{w}$.

The vector $3\mathbf{v} = \mathbf{v} + \mathbf{v} + \mathbf{v}$ has the same direction as $\mathbf{v}$ but is 3 times longer. Generalizing this example one arrives at the following definition of the multiplication of a vector by a scalar: given a vector $\mathbf{v}$ and a real number α, the result of their multiplication is a vector, denoted $\alpha\mathbf{v}$, which has the same direction as $\mathbf{v}$ but is α times longer. The last phrase calls for comments since it is literally true only for $\alpha > 1$. If $0 < \alpha < 1$, being "α times longer" actually means "shorter". If $\alpha < 0$, the direction of α is in fact opposite to the direction of $\mathbf{v}$. Finally, $0\mathbf{v} = \mathbf{0}$ is the zero vector represented by directed segments $\vec{AA}$ of zero length.

Combining the operations of vector addition and multiplication by scalars we can form expressions $\alpha\mathbf{u} + \beta\mathbf{v} + ... + \gamma\mathbf{w}$ which are called linear combinations of vectors $\mathbf{u}, \mathbf{v}, ..., \mathbf{w}$ with coefficients $\alpha, \beta, ..., \gamma$. Linear combinations will regularly occur throughout the course.

1.1.2. Inner product. Metric concepts of elementary Euclidean geometry, such as lengths and angles, can be conveniently encoded by the operation of inner product of vectors (also known as scalar product or dot-product). Given two vectors $\mathbf{u}$ and $\mathbf{v}$ of lengths $|\mathbf{u}|$ and $|\mathbf{v}|$ and making the angle θ to each other, their inner product is a *number* defined by the formula:

$$\langle \mathbf{u}, \mathbf{v} \rangle = |\mathbf{u}|\ |\mathbf{v}|\ \cos\theta.$$

It is clear from the definition that

(a) the inner product is symmetric: $\langle \mathbf{u}, \mathbf{v} \rangle = \langle \mathbf{v}, \mathbf{u} \rangle$,

(b) non-zero vectors have positive inner squares $\langle \mathbf{u}, \mathbf{u} \rangle = |\mathbf{u}|^2$

(c) the angle θ is recovered form the inner products via

$$\cos\theta = \frac{\langle \mathbf{u}, \mathbf{v} \rangle}{\langle \mathbf{u}, \mathbf{u} \rangle^{1/2} \langle \mathbf{v}, \mathbf{v} \rangle^{1/2}}.$$

We will see soon that (even though it is not obvious from the definition) the inner product also has the following nice algebraic properties called bilinearity:

$$\langle \alpha\mathbf{u} + \beta\mathbf{v}, \mathbf{w} \rangle = \alpha\langle \mathbf{u}, \mathbf{w} \rangle + \beta\langle \mathbf{v}, \mathbf{w} \rangle$$

$$\langle \mathbf{w}, \alpha\mathbf{u} + \beta\mathbf{v} \rangle = \alpha\langle \mathbf{w}, \mathbf{u} \rangle + \beta\langle \mathbf{w}, \mathbf{v} \rangle.$$

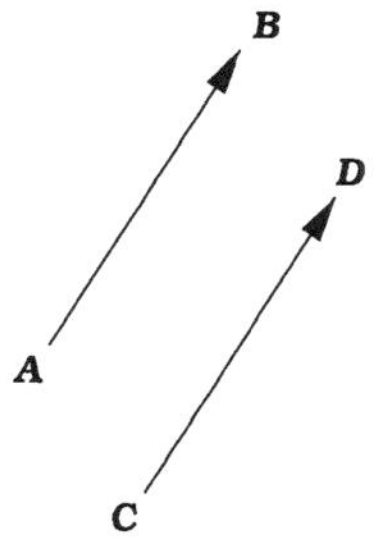

Exercises 1.1.1.

(a) A mass m rests on an reclined plane making the angle $\pi/6$ to the horizontal direction. Find the forces of friction and reaction by which the surface acts on the mass.

(b) A ferry capable of making 5 mph shuttles across a river of width 0.2 mi with a strong current of 3 mph. How long does each round trip take?

(c) Let $ABCD$ be a parallelogram. Prove the vector equality $\vec{AC} = \vec{AB} + \vec{AD}$ and derive the **commutativity** of the vector sum: $\mathbf{u} + \mathbf{v} = \mathbf{v} + \mathbf{u}$.

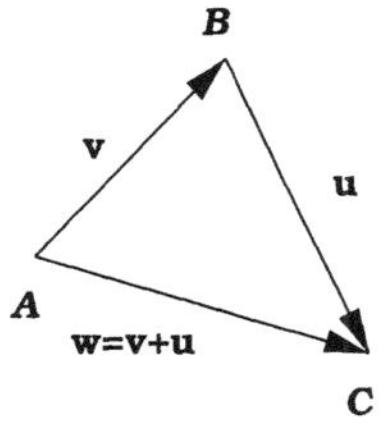

(d) Examine the picture that looks like the projection of a $3D$-cube to the plane and prove **associativity** of the vector sum: $(\mathbf{u} + \mathbf{v}) + \mathbf{w} = \mathbf{u} + (\mathbf{v} + \mathbf{w})$.

(e) Three medians of a triangle ABC intersect at one point M called the **barycenter** of the triangle. Let O be any point on the plane. Prove the vector equality

$$\vec{OM} = \frac{1}{3}(\vec{OA} + \vec{OB} + \vec{OC}).$$

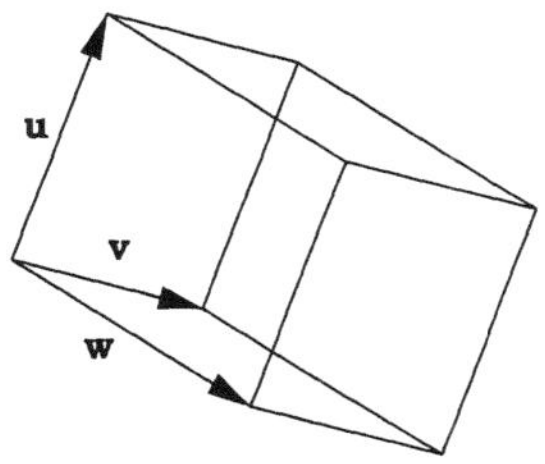

(f) Three points A, B, C revolve clockwise with the same angular velocity along three circles (of possibly different radii) centered at there different points O_A,O_B,O_C. Show that the trajectory of the barycenter of the triangle ABC is a circle and find its center.

(g) Given a triangle ABC, we construct a new triangle $A'B'C'$ in such a way that A' is centrally symmetric to A with respect to the center B, B' — symmetric to B with respect to C, C' — symmetric to C with respect to A, and then erase the original triangle. Reconstruct ABC from $A'B'C'$ by straightedge and compass.

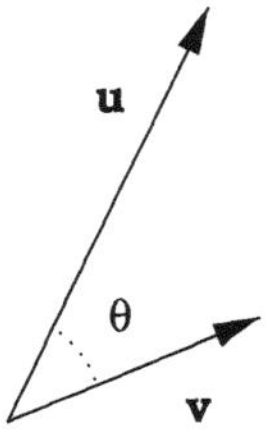

Exercises 1.1.2.

(a) Prove the **Cauchy – Schwartz** inequality $\langle \mathbf{u}, \mathbf{v}\rangle^2 \leq \langle \mathbf{u}, \mathbf{u}\rangle\langle \mathbf{v}, \mathbf{v}\rangle$. In which cases does the inequality turn into equality? Deduce the **triangle inequality**

$$|\mathbf{u} + \mathbf{v}| \leq |\mathbf{u}| + |\mathbf{v}|.$$

(b) Compute the inner product $\langle \vec{AB}, \vec{BC}\rangle$ if ABC is a regular triangle inscribed into a unit circle.

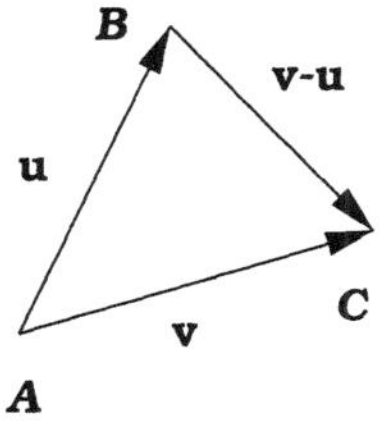

(c) Express the inner product $\langle \mathbf{u}, \mathbf{v}\rangle$ in terms of the lengths $|\mathbf{u}|, |\mathbf{v}|, |\mathbf{u} + \mathbf{v}|$ of the two vectors and of their sum.

Example. Given a triangle ABC, let us denote by $\mathbf{u}$ and $\mathbf{v}$ the vectors represented by the directed segments $\vec{AB}$ and $\vec{AC}$ and use properties of the inner product in order to compute the length $|BC|$. Notice that the segment $\vec{BC}$ represents $\mathbf{v}-\mathbf{u}$. We have:

$$|BC|^2 = \langle \mathbf{v}-\mathbf{u}, \mathbf{v}-\mathbf{u}\rangle = \langle \mathbf{v}, \mathbf{v}\rangle + \langle \mathbf{u}, \mathbf{u}\rangle - 2\langle \mathbf{u}, \mathbf{v}\rangle$$
$$= |AC|^2 + |AB|^2 - 2|AB|\,|AC|\,\cos\theta.$$

This is the famous "cosine theorem" from trigonometry. If the vectors $\mathbf{u}$ and $\mathbf{v}$ are perpendicular, that is $\theta = \pi/2$, then $\langle \mathbf{u}, \mathbf{v}\rangle = 0$, and the cosine formula turns into the Pythagorean theorem.

Vectors with the inner product $\langle \mathbf{u}, \mathbf{v}\rangle = 0$ are called orthogonal. Orthogonality of vectors means that either the angle between the vectors is $\pi/2$ or at least one of the vectors is zero (in which case the angle is not defined).

1.1.3. Coordinates. One introduces a Cartesian coordinate system on the Euclidean plane by choosing the origin O and specifying directions of two perpendicular coordinate axes. A point U on the plane determines the vector $\vec{OU}$ called the radius-vector of the point with respect to the origin. Vice versa, any vector $\mathbf{u}$ on the plane can be represented by a directed segment $\vec{OU}$ with the tail O. Thus $\mathbf{u}$ is unambiguously specified by the coordinates (u_1, u_2) of the head U which are called the coordinates of the vector. According to a convention that has become standard in mathematics (and is very awkward for typesetting), vectors are written by columns of their coordinates: $\mathbf{u} = \begin{bmatrix} u_1 \\ u_2 \end{bmatrix}$. Notice that the same vector may have different coordinates in a different coordinate system.

The operations of vector sum, multiplication by scalars and the inner product have the following simple coordinate expressions:

$$\mathbf{u}+\mathbf{v} = \begin{bmatrix} u_1 & + & v_1 \\ u_2 & + & v_2 \end{bmatrix}, \quad \alpha\mathbf{u} = \begin{bmatrix} \alpha u_1 \\ \alpha u_2 \end{bmatrix}, \quad \langle \mathbf{u}, \mathbf{v}\rangle = u_1v_1 + u_2v_2.$$

The first formula means that coordinates of the vertex W of the parallelogram $OUWV$ are sums of the corresponding coordinates of U and V. This should be clear from congruence of the triangles OUU' and VWW' on the picture. The second formula follows from similarity of the triangles OUU' and OVV'. In order to prove the coordinate formula for the inner product, we denote by ϕ and ψ the angles the vectors $\mathbf{u}$ and $\mathbf{v}$ make with the positive direction of the 1-st axis and use $u_1 = |\mathbf{u}|\cos\phi, u_2 = |\mathbf{u}|\sin\phi, v_1 = |\mathbf{v}|\cos\psi, v_2 = |\mathbf{v}|\sin\psi$. We find

$$u_1v_1 + u_2v_2 = |\mathbf{u}|\,|\mathbf{v}|(\cos\phi\cos\psi + \sin\phi\sin\psi) = |\mathbf{u}|\,|\mathbf{v}|\,\cos(\phi-\psi)$$

due to the addition formula in trigonometry. It remains to notice that the angle θ between $\mathbf{v}$ and $\mathbf{u}$ equals $\phi - \psi$.

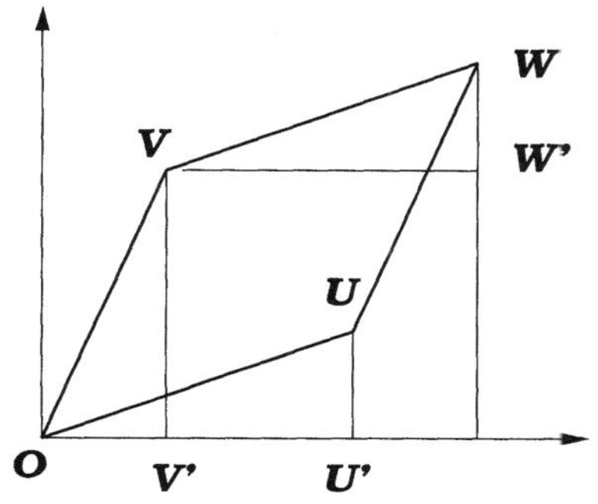
W
V
W'
U
O
V'
U'

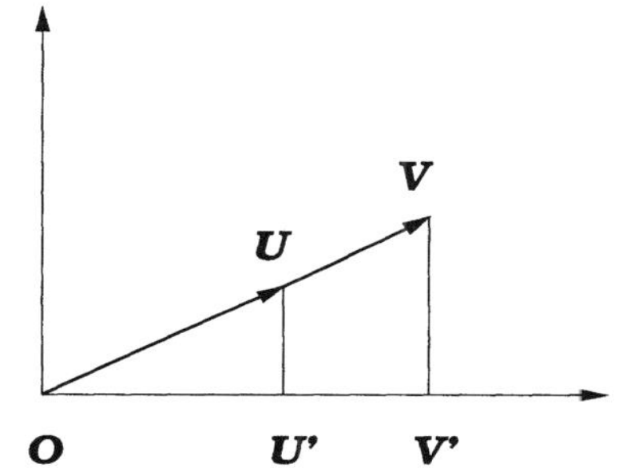
V
U
O
U'
V'

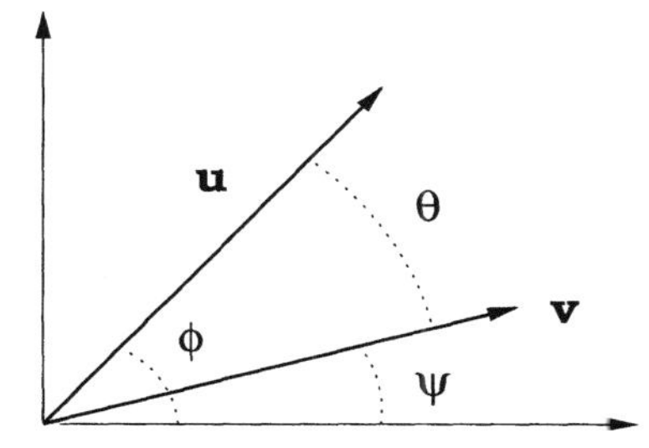
u
θ
v
ϕ
ψ

Now we are ready to check the bilinearity property of the inner product by a straightforward computation:

$$\langle \alpha\mathbf{u}, \mathbf{w}\rangle = (\alpha u_1)w_1 + (\alpha u_2)w_2 = \alpha(u_1w_1 + u_2w_2) = \alpha\langle \mathbf{u}, \mathbf{w}\rangle,$$

$$\langle \mathbf{u}+\mathbf{v}, \mathbf{w}\rangle = (u_1+v_1)w_1 + (u_2+v_2)w_2 = (u_1w_1+u_2w_2) + (v_1w_1+v_2w_2)$$

$$= \langle \mathbf{u}, \mathbf{w}\rangle + \langle \mathbf{v}, \mathbf{w}\rangle.$$

The unit coordinate vectors $\mathbf{e}_1 = \begin{bmatrix} 1 \\ 0 \end{bmatrix}$, $\mathbf{e}_2 = \begin{bmatrix} 0 \\ 1 \end{bmatrix}$ are said to form an orthonormal basis of the plane. The term "orthonormal" refers to the orthogonality of the vectors ($\langle \mathbf{e}_1, \mathbf{e}_2\rangle = 0$) and their unit length ($\langle \mathbf{e}_1, \mathbf{e}_1\rangle = \langle \mathbf{e}_2, \mathbf{e}_2\rangle = 1$) while "basis" refers to the following property: any vector $\mathbf{u}$ is uniquely written as a linear combination of $\mathbf{e}_1, \mathbf{e}_2$,

$$\mathbf{u} = u_1\mathbf{e}_1 + u_2\mathbf{e}_2 = u_1\begin{bmatrix} 1 \\ 0 \end{bmatrix} + u_2\begin{bmatrix} 0 \\ 1 \end{bmatrix} = \begin{bmatrix} u_1 \\ u_2 \end{bmatrix}.$$

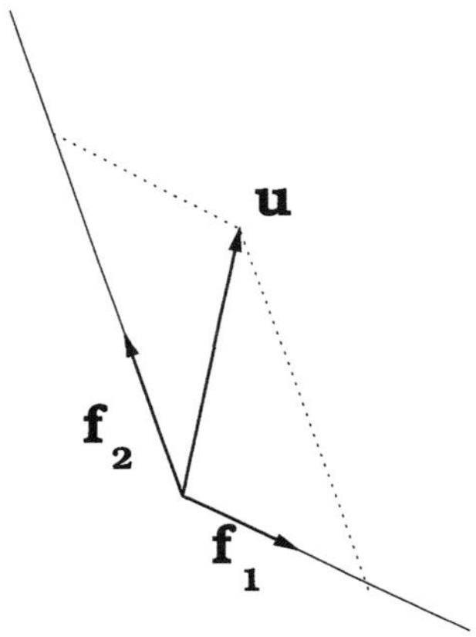

More general coordinate systems on the plane are obtained by picking any two non-proportional vectors $\mathbf{f}_1, \mathbf{f}_2$ on the role of a basis. Given such $\mathbf{f}_1, \mathbf{f}_2$, any vector $\mathbf{u}$ is uniquely written as a linear combination $\mathbf{u} = u_1\mathbf{f}_1 + u_2\mathbf{f}_2$ (the picture shows how). The column $\begin{bmatrix} u_1 \\ u_2 \end{bmatrix}$ is taken on the role of coordinates of $\mathbf{u}$ in the basis $\mathbf{f}_1, \mathbf{f}_2$. Addition of vectors and multiplication by scalars is described by the same component-wise formulas as before (say, $\mathbf{w} = \mathbf{u}+\mathbf{v} = (u_1\mathbf{f}_1 + u_2\mathbf{f}_2) + (v_1\mathbf{f}_1 + v_2\mathbf{f}_2) = (u_1+v_1)\mathbf{f}_1 + (u_2+v_2)\mathbf{f}_2$ which means that $w_1 = u_1 + v_1, w_2 = u_2 + v_2$). However, the inner product formula takes on a more general form:

$$\langle \mathbf{u}, \mathbf{v}\rangle = au_1v_1 + bu_1v_2 + bu_2v_1 + cu_2v_2,$$

where $a = \langle \mathbf{f}_1, \mathbf{f}_1\rangle, b = \langle \mathbf{f}_1, \mathbf{f}_2\rangle, c = \langle \mathbf{f}_2, \mathbf{f}_2\rangle$.

Exercises 1.1.3.

(a) Check the formula for the inner product in a general basis $\mathbf{f}_1, \mathbf{f}_2$.

(b) Let $A_1A_2A_3A_4A_5A_6$ be the regular hexagon inscribed into the unit circle centered at the origin O and such that $\vec{OA_1} = \mathbf{e}_1$. Find coordinates of all vertices of the hexagon. Take $\mathbf{f}_1 = \vec{OA_2}$ and $\mathbf{f}_2 = \vec{OA_6}$ on the role of a new basis and compute coordinates of all the vertices with respect to this basis. Find the coordinate formula for the inner product in the basis $\mathbf{f}_1, \mathbf{f}_2$.

(c) Let $\begin{bmatrix} a \\ b \end{bmatrix}$ and $\begin{bmatrix} c \\ d \end{bmatrix}$ be Cartesian coordinates of A and B. Prove that the area of the triangle OAB equals the absolute value of $(ad - bc)/2$.

1.2. Analytical geometry

Multivariable Calculus of functions in two variables relies on basic properties of linear and quadratic functions studied in Analytical Geometry and Linear Algebra.

1.2.1. Linear functions and straight lines. The trajectory of a point moving with a constant velocity $\mathbf{v}$ is a straight line. The radius-vector of the point at the moment t is $\mathbf{x}(t) = \mathbf{x}(0) + t\mathbf{v}$ where $\mathbf{x}(0)$ specifies the initial position of the point at $t = 0$. This is the vector form of the following parametric equations of the line:

$$\begin{array}{rcrcl} x_1(t) & = & x_1(0) & + & v_1 t \\ x_2(t) & = & x_2(0) & + & v_2 t \end{array}.$$

Eliminating t from these formulas we arrive at a (non-parametric) equation of the line which has the form

$$a_1x_1 + a_2x_2 = b$$

where at least one of the coefficients a_1, a_2 is non-zero. The left hand side of this equation is a linear function (or linear form) of the two variables (x_1, x_2):

$$y = a_1x_1 + a_2x_2.$$

The graph of a function f in two variables is the surface in the 3-space given by the equation $y = f(x_1, x_2)$. The graph of a linear form is a plane passing through the origin $(x_1, x_2, y) = (0, 0, 0)$. The curves on the plane where a given function of two variables is constant, are called levels or level curves of the function. The zero level of the linear form is the line $a_1x_1 + a_2x_2 = 0$ passing through the origin, and the other levels are lines parallel to it.

Exercises 1.2.1.

(a) Recall the definition of parallel lines and prove the last statement of 1.2.1.

(b) Find all equations of the form $a_1x_1 + a_2x_2 = b$ which describe a line passing through the origin.

(c) Which equations of the form $a_1x_1 + a_2x_2 = b$ describe the same lines?

(d) Prove that the lines $ax_1 + bx_2 = e$ and $cx_1 + dx_2 = f$ intersect at one point if and only if $ad - bc \neq 0$.

(e) Two ships cruising in the ocean with velocity vectors $(8, 6)$ and $(-6, 8)$ are located at the points $(0, 0)$ and $(100, 0)$ at the moment $t = 0$. Find the intersection point of their trajectories. Will the ships collide?

(f) A friend of mine working as a manager in a leading company in Silicon Valley claims that he is the only person around him who knows how to write an equations of the line passing through two points with given coordinates. Prove that you deserve a manager position in Silicon Valley: derive the damn thing!

1.2.2. Conic sections. According to Kepler's law of planetary motion, planets, asteroids and comets revolve along elliptic orbits with the Sun at one of the foci. Ellipses are examples of conic sections — plane curves introduced by ancient Greeks as intersections of the plane with a circular conic surface in the 3-space. The picture shows how to locate the two foci of the ellipse: place into the conic "cup" two balls of such sizes that they touch the cutting plane from opposite sides. Then the tangency points of the balls with the plane are the foci of the ellipse. This picture also demonstrates the ancient proof of the fact that the ellipse consists of all points on the plane with a fixed sum of distances to the foci.

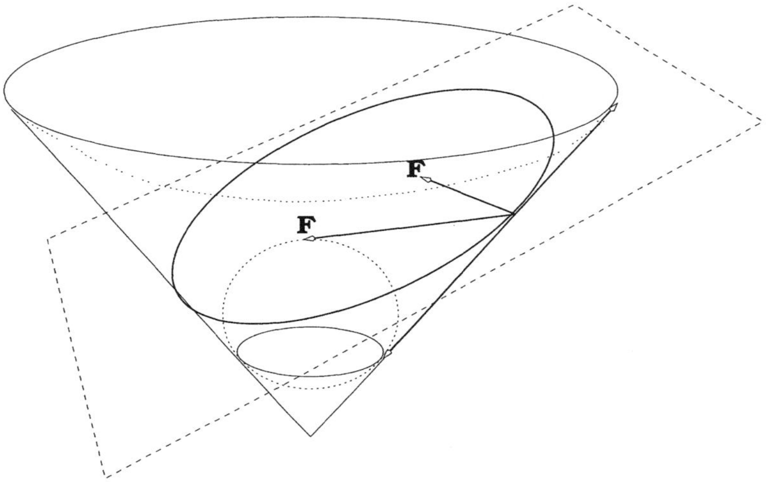

F - *foci of the conical section are the tangency points of the secting plane with the balls*

From the viewpoint of Descartes' analytical geometry conic sections are plane curves given by quadratic equations

$$ax_1^2 + 2bx_1x_2 + cx_2^2 + dx_1 + ex_2 + f = 0.$$

Indeed, the circular cone in the 3 space can be described by the equation $x_1^2 + x_2^2 = x_3^2$. Substituting the equation $x_3 = \alpha_1 x_1 + \alpha_2 x_2 + \beta$ of the cutting plane we obtain a quadratic relation among x_1 and x_2 (which in fact describes the projection of the conic section to the horizontal plane). We list model examples of quadratic curves.

Examples. (a) $x_1^2 + x_2^2 = 1$ is the unit circle centered at the origin.

(b) The normal ellipse with semiaxes $a_1 > a_2 > 0$ is described by the equation

$$\frac{x_1^2}{a_1^2} + \frac{x_2^2}{a_2^2} = 1.$$

The ellipse intersects the coordinate axes at the points $(\pm a_1, 0)$ and $(0, \pm a_2)$ and is obtained from the unit circle by stretching a_1 times in the direction of x_1 and a_2 times in the direction of x_2. The coordinate lines are the symmetry axes of the normal ellipse and are often called its principal axes.

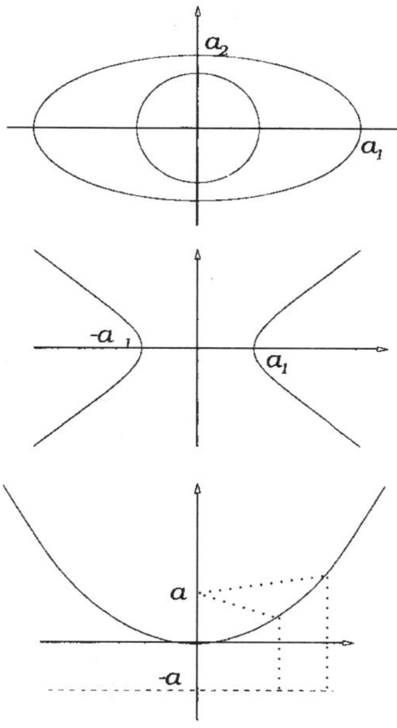

Normal ellipses, hyperbolas and parabolas

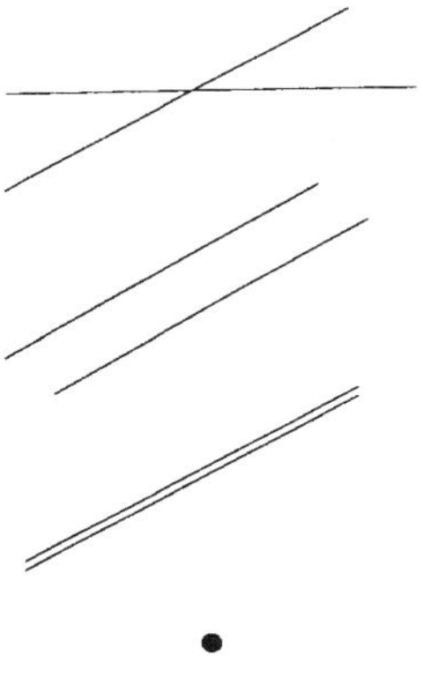

Degenerate Curves

Exercises 1.2.2.

(a) Sketch the ellipses

$$x_1^2 + 4x_2^2 = 1,$$

$$x_1^2 + x_2^2/4 = 1,$$

$$x_1^2 + x_2^2/4 = 4$$

"to scale".

(b) Sketch the level curves

$$4x_1^2 - x_2^2 = 1,\ 0,\ -1.$$

(c) Find the area enclosed by the ellipse with semi-axes a and b.

(d) Using the fact that all tangent segments from a given point to a sphere have equal lengths, prove that an elliptic conic section consists of points on the intersecting plane with a constant sum of distances to the foci.

(e) Figure out which sections of the circular cone are hyperbolas and parabolas. Prove that the hyperbolas consist of points on the intersecting plane with a constant difference of distances to the foci.

(f) Prove that the parabola $y = x^2/4a$ consists of points equidistant to the focus $(x, y) = (0, a)$ and to the line $y = -a$ called the directrix.

(g) Compute coordinates of the foci for the normal ellipses and hyperbolas.

(h) Quadratic curves have special "optical" properties. Show that the light rays originating from a focus of an ellipse and reflected in the ellipse as in the mirror will focus at the other focus. Formulate and prove corresponding properties of hyperbolas and parabolas.

(c) The curve $xy = 1$ is the graph of the function $y = 1/x$ and is called a hyperbola. Changing the coordinates by $x = (x_1 - x_2), y = (x_1 + x_2)$ we transform the equation to $x_1^2 - x_2^2 = 1$. Stretching in the directions of x_1 and x_2 with coefficients a_1 and a_2 produce the normal hyperbola with "semiaxes a_1, a_2":

$$\frac{x_1^2}{a_1^2} - \frac{x_2^2}{a_2^2} = 1.$$

It is symmetric with respect to the coordinate axes.

(d) The normal parabolas are given by the equations $x_2 = x_1^2/4a$.

(e) Some quadratic equations are degenerate and describe two straight lines (intersecting, parallel or even coinciding), or a point, or the whole plane, or the empty set.

We will show that any *non*-degenerate quadratic curve is either an ellipse, or hyperbola, or parabola and is given by one of the normal equations in a suitable Cartesian coordinate system on the plane.

1.2.3. Quadratic forms. The quadratic function on the left hand side of the quadratic equation is the sum of a constant, a linear form and a quadratic form, $ax_1^2 + 2bx_1x_2 + cx_2^2$. The classification of quadratic curves is based on the following theorem about quadratic forms. A line through the origin is called a principal axis of a given quadratic form, if the quadratic form is symmetric with respect to this line, that is takes on the same values at any two symmetric points. For example, changes of signs of x_1 or x_2 do not change the quadratic form $ax_1^2 + cx_2^2$, which shows that the coordinate axes are principal for this quadratic form. The inner square $\langle \mathbf{x}, \mathbf{x} \rangle = x_1^2 + x_2^2$ is symmetric about any line through the origin and thus has infinitely many principal axes.

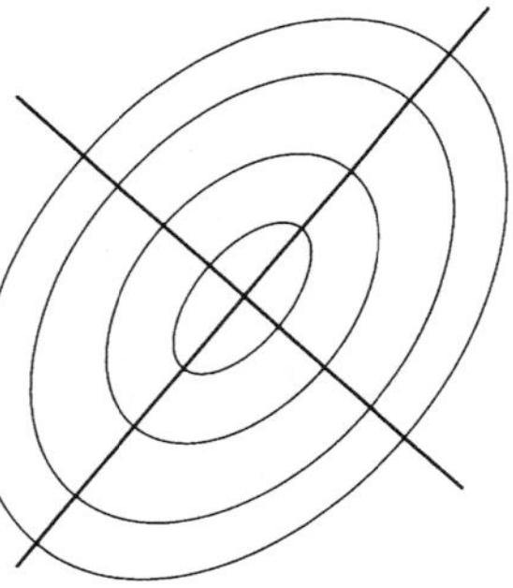

Principal axes and level curves

Theorem (Principal Axis Theorem).
Any quadratic form in two variables has two perpendicular principal axes.

As we have just seen, the theorem follows from

Theorem $'$ (Orthogonal Diagonalization Theorem).
Any quadratic form in a suitable Cartesian coordinate system takes on $AX_1^2 + CX_2^2$.

In its turn Orthogonal Diagonalization Theorem follows from Principal Axis Theorem so that they are actually equivalent. Indeed, choose an orthonormal basis $\mathbf{f}_1, \mathbf{f}_2$ in such a way that $\mathbf{f}_1$ has a principal direction for the quadratic form in question. Written in the new coordinate system as $AX_1^2 + 2BX_1X_2 + CX_2^2$, the quadratic form is symmetric with respect to the change $X_1 \mapsto -X_1$ by the definition of a principal axis. Thus $B = 0$.

By the way, this argument shows that the axis perpendicular to a principal one is principal automatically.

Examples. (a) The quadratic form xy is symmetric about the line $x = y$. Take $\mathbf{f}_1 = (1/\sqrt{2}, 1/\sqrt{2})$ and the perpendicular unit vector $\mathbf{f}_2 = (1/\sqrt{2}, -1/\sqrt{2})$ on the role of the new orthonormal basis. In the new coordinates $X_1 = (x+y)/\sqrt{2}$ and $X_2 = (x-y)/\sqrt{2}$ the quadratic form xy reads $X_1^2/2 - X_2^2/2$.

The next two examples show how to bring equations of quadratic curves to one of the normal forms assuming that the quadratic part of the equation has been already transformed to principal axes by a rotation of the coordinate axes and is therefore equal to $ax_1^2 + cx_2^2$.

(b) If both coefficients a, c in the equation $ax_1^2 + cx_2^2 + dx_1 + ex_2 + f = 0$ are zeroes, the equation is not really quadratic and determines a straight line. If, say, $c = 0$ but $a \neq 0$, we can transform the equation to $aX_1^2 + ex_2 + F = 0$ by "completing the square" $a(x_1 + d/2a)^2 - d^2/4a + ex_2 + f$ and shifting the origin: $X_1 = x_1 + d/2a$. If $e = 0$, the resulting equation defines a degenerate curve — two parallel lines (when a and F have opposite signs), or the empty set (when the sign is the same), or a double line (when $F = 0$). If $e \neq 0$, we arrive at one of the normal equations $X_2 = -aX_1^2/e$ for parabolas by putting $X_2 = x_2 + F/e$.

(c) If both a, c are non-zero, we transform the equation to the form $aX_1^2 + cX_2^2 + F = 0$ by completing squares in each of the variables. If $F = 0$ the equation defines one point (when a and c have the same sign) or a pair of lines meeting at the origin (when the signs are opposite). If $F \neq 0$, the curve is either one of the normal ellipses and hyperbolas, depending on the signs of a and c, or it is empty (when $a, c, F > 0$).

We see that transformations of quadratic equations to the normal forms involve rotations of coordinate systems, shift of the origin and division of the equation by a non-zero constant. It remains unclear at this point how to find the rotated coordinate system in which a given quadratic form takes on $AX_1^2 + CX_2^2$. A routine procedure for doing this is contained in the proof of the Orthogonal Diagonalization Theorem which we postpone till 1.5.2. Meanwhile we give another application of the theorem: by rescaling the coordinates X_1, X_2 we make the coefficients A, C equal to $+1, -1$ or 0 and arrive at the following classification of quadratic forms up to change of variables:

Corollary. *Any quadratic form in a suitable (not necessarily Cartesian) coordinate system assumes one of the* normal forms*:*

$$X_1^2 + X_2^2, \; X_1^2 - X_2^2, \; -X_1^2 - X_2^2, \; X_1^2, \; -X_2^2, \; 0.$$

Exercises 1.2.3.

(a) Using the symmetry of the following quadratic forms about the line $x = y$, transform them to principal axes and find semiaxes of the corresponding curve $Q(x, y) = 1$ if the curve is an ellipse or hyperbola:

$$Q = x^2 - xy + y^2, \quad Q = x^2 - 2xy + y^2, \quad Q = x^2 - 4xy + y^2.$$

(b) Find the place of the quadratic curve $x_1^2 - 4x_2^2 = x_1 - 4x_2$ in the classification.

(c) Sketch the graphs of the functions $(x_1^2 + x_2^2)$ (paraboloid) and $x_1^2 - x_2^2$ (saddle).

(d) Study the sign of the quadratic form $ax_1^2 + 2bx_1x_2 + cx_2^2$ restricted to the lines $x_2 = kx_1$ through the origin. Show that the quadratic form has the minimum at the origin if $ac - b^2 > 0$ and $a, c > 0$, the maximum — if $ac - b^2 > 0$ and $a, c < 0$, and has no minimum/maximum if $ac - b^2 < 0$. What happens if $ac - b^2 = 0$? Deduce that the sign of $ac - b^2$ does not depend on the choice of the coordinate system.

(e) Which of the following curves are ellipses and which are hyperbolas?

$$x_1^2 + 4x_1x_2 = 1, \quad x_1^2 + 2x_1x_2 + 4x_2^2 = 1, \quad x_1^2 + 4x_1x_2 + 4x_2^2 = 1, \quad x_1^2 + 6x_1x_2 + 4x_2^2 = 1$$

1.3. Linear transformations and matrices

Reflections about a point, about a line, stretching along coordinate axes we encountered recently are examples of linear transformations on the plane.

1.3.1. Linearity. The definition of linear transformations fits the abstract concept of a function from a set X to a set Y: a function is a rule that to each element of the set X associates exactly one element of the set Y. A linear transformation T on the plane is a rule that to each vector $\mathbf{x}$ associates the vector $T\mathbf{x}$ on the same plane in such a way that linear combinations of any vectors $\mathbf{x}, \mathbf{y}$ with any coefficients α, β are transformed to linear combinations with the same coefficients:

$$T(\alpha\mathbf{x} + \beta\mathbf{y}) = \alpha T\mathbf{x} + \beta T\mathbf{y}.$$

This property is called linearity and means geometrically that proportional vectors are transformed to proportional vectors (in particular, the origin is preserved: $T\mathbf{0} = \mathbf{0}$), and parallelograms are transformed to parallelograms.

Example. (a) The rotation about the origin is a linear transformation since it transforms parallelograms to parallelograms and proportional vectors — to proportional ones. Similarly, the simultaneous reflection of all vectors about a line passing through the origin is a linear transformation too. Notice that these linear transformations preserve lengths of all vectors and angles between them and thus preserve the inner product:

$$\langle T\mathbf{x}, T\mathbf{y}\rangle = \langle \mathbf{x}, \mathbf{y}\rangle \text{ for all } \mathbf{x}, \mathbf{y}.$$

Linear transformations which preserve inner products of all vectors are called orthogonal transformations. We will see soon that any orthogonal transformation on the plane is either a rotation about the origin or the reflection about a line through the origin. Notice that reflections reverse the clockwise direction of the unit circle to the counterclockwise direction, while rotations preserve the directions.

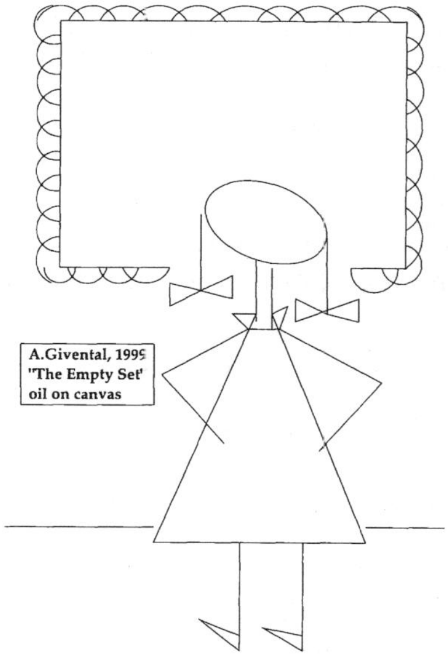

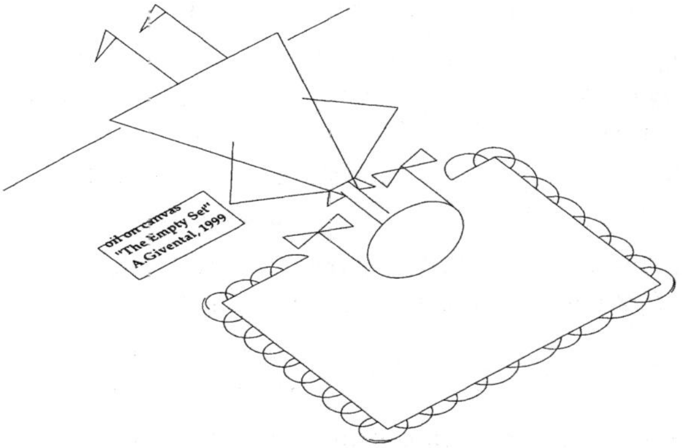

Is this transformation linear?

Any linear transformation is uniquely determined by the way it acts on a basis. Indeed, if $\mathbf{f}_1, \mathbf{f}_2$ is a basis, and the vectors $T\mathbf{f}_1$, $T\mathbf{f}_2$ are given, the vector $T\mathbf{x} = T(x_1\mathbf{f}_1 + x_2\mathbf{f}_2) = x_1 T\mathbf{f}_1 + x_2 T\mathbf{f}_2$ is determined by the coordinates (x_1, x_2) of the vector $\mathbf{x}$. Vice versa, if $\mathbf{f}_1, \mathbf{f}_2$ is a basis, we can pick any two vectors $\mathbf{v}_1, \mathbf{v}_2$ on the role of $T\mathbf{f}_1$ and $T\mathbf{f}_2$ and then extend the rule T to all vectors by the formula $T\mathbf{x} = x_1\mathbf{v}_1 + x_2\mathbf{v}_2$. It is not hard to check (do it!) that the transformation T defined by this rule satisfies the linearity condition.

In coordinates, the linear transformation is given by two linear forms:

$$\begin{array}{ccccc} x_1' & = & ax_1 & + & bx_2 \\ x_2' & = & cx_1 & + & dx_2 \end{array}.$$

The columns $\begin{bmatrix} a \\ c \end{bmatrix}$ and $\begin{bmatrix} b \\ d \end{bmatrix}$ here represent the vectors $T\mathbf{f}_1$ and $T\mathbf{f}_2$ in the basis $\mathbf{f}_1, \mathbf{f}_2$, and $\begin{bmatrix} x_1' \\ x_2' \end{bmatrix}$ are the coordinates of $\mathbf{x}' = T\mathbf{x}$. A standard notational convention suggests to combine the columns into a single 2×2-matrix of the linear transformation T with respect to the basis:

$$T = \begin{bmatrix} a & b \\ c & d \end{bmatrix}.$$

We will usually denote the matrix of a linear transformation by the same capital letter as the linear transformation itself. (This notation assumes however that it is clear from the context what basis we have in mind, since the same linear transformation may have different matrices in different bases.)

Examples. (b) The rotation T_θ through the angle θ (counted counterclockwise) transforms the orthonormal basis $\mathbf{e}_1, \mathbf{e}_2$ to $(\cos\theta)\mathbf{e}_1 + (\sin\theta)\mathbf{e}_2$, $-(\sin\theta)\mathbf{e}_1 + (\cos\theta)\mathbf{e}_2$. Thus the matrix of the rotation is

$$T_\theta = \begin{bmatrix} \cos\theta & -\sin\theta \\ \sin\theta & \cos\theta \end{bmatrix}.$$

(c) A similar computation shows that the reflection R_θ about the axis making the angle $\theta/2$ with the positive direction of the x_1-axis has the matrix

$$R_\theta = \begin{bmatrix} \cos\theta & \sin\theta \\ \sin\theta & -\cos\theta \end{bmatrix}.$$

(d) Any orthogonal transformation T transforms the orthonormal basis $\mathbf{e}_1, \mathbf{e}_2$ to the basis $T\mathbf{e}_1, T\mathbf{e}_2$ which must be orthonormal too. A unit vector $T\mathbf{e}_1$ is written as $(\cos\theta)\mathbf{e}_1 + (\sin\theta)\mathbf{e}_2$ for a suitable θ. The choice of $T\mathbf{e}_1$ leaves two (opposite) choices for the unit vector $T\mathbf{e}_2$ perpendicular to $T\mathbf{e}_1$: $\pm((\sin\theta)\mathbf{e}_1 - (\cos\theta)\mathbf{e}_2)$. One of the choices gives rise to the reflection matrix, the other — to the rotation matrix. Thus, any orthogonal transformation on the plane is either rotation or reflection.

1.3.2. Composition. Composition of abstract functions $g : X \to Y$ and $f : Y \to Z$ is defined as a function $h : X \to Z$ by consecutive application of the two rules: $h(x) = f(g(x))$. Composition of two linear transformations from the plane to itself is a linear transformation too:

$$A(B(\alpha\mathbf{x} + \beta\mathbf{y})) = A(\alpha B(\mathbf{x}) + \beta B(\mathbf{y})) = \alpha A(B(\mathbf{x})) + \beta A(B(\mathbf{y})).$$

Exercises 1.3.1.

(a) Is SID a function? and SSN?

(b) Describe geometrically the linear transformations defined by the 8 matrices

$$\begin{bmatrix} \pm 1 & 0 \\ 0 & \pm 1 \end{bmatrix}, \begin{bmatrix} 0 & \pm 1 \\ \pm 1 & 0 \end{bmatrix}.$$

(c) Find the matrix of a linear transformation which transforms the hyperbola $x_1x_2 = 1$ to $x_1^2 - x_2^2 = 1$.

(d) Find matrices of all linear transformation which transform the unit square

$$0 \leq x_1, x_2 \leq 1$$

to the parallelogram

$$0 \leq x_1 + x_2, x_2 \leq 1.$$

(e) Find 6 linear transformations which preserve the equilateral triangle ABC centered at the origin and with $\vec{OA} = \mathbf{e}_1$. Find matrices of these linear transformations in the basis $\mathbf{e}_1, \mathbf{e}_2$ and in the basis $\vec{OB}, \vec{OC}$.

(f) Which of the transformations from Exercises (a – d) are orthogonal? rotations? reflections?

(g) Prove that any linear transformation which preserves length of all vectors also preserves angles between any vectors and is therefore orthogonal.

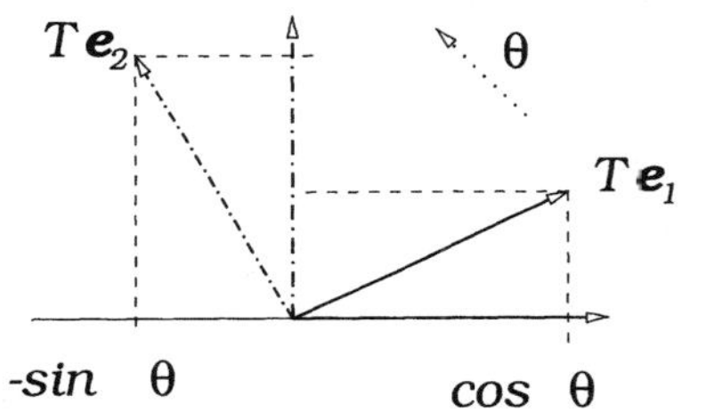

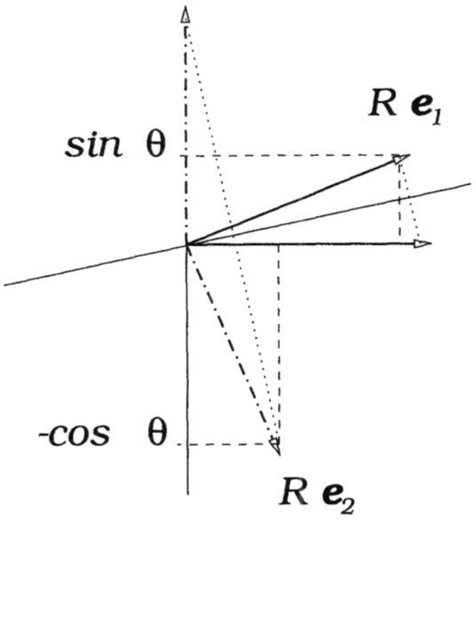

Example. (a) The composition of two different reflections is a rotation. Indeed, it preserves inner products (since each reflection does) and restores the direction of the unit circle (since each reflection reverses it). Compositions of two rotations are rotations, and reflections composed with rotations in any order give rise to reflections (why?)

Composing linear transformations

$$\begin{array}{l} x_1'' = a_{11}x_1' + a_{12}x_2' \\ x_2'' = a_{21}x_1' + a_{22}x_2' \end{array}, \quad \begin{array}{l} x_1' = b_{11}x_1 + b_{12}x_2 \\ x_2' = b_{21}x_1 + b_{22}x_2 \end{array}$$

we arrive, after a straightforward computation, to a linear transformation with the matrix

$$\begin{bmatrix} c_{11} & c_{12} \\ c_{21} & c_{22} \end{bmatrix} = \begin{bmatrix} a_{11}b_{11} + a_{12}b_{21} & a_{11}b_{12} + a_{12}b_{22} \\ a_{21}b_{11} + a_{22}b_{21} & a_{21}b_{12} + a_{22}b_{22} \end{bmatrix}.$$

The matrix C is called the **matrix product** of the matrices A and B and is denoted AB. In order to discover a pattern in the above formulas, let us introduce the **product of a row** $[a_1, a_2]$ **with a column** $\begin{bmatrix} b_1 \\ b_2 \end{bmatrix}$ to be a number defined as $a_1b_1 + a_2b_2$. The subscripts of the matrix entries c_{ij} of a matrix C specify the row i and the column j where the entry is located. In these notations the entry c_{ij} of the product matrix $C = AB$ equals the product of the i-th row of A with the j-th column of B.

Example. (b) The product DA (respectively AD) of the **diagonal** matrix $D = \begin{bmatrix} d_1 & 0 \\ 0 & d_2 \end{bmatrix}$ with $A = \begin{bmatrix} a_{11} & a_{12} \\ a_{21} & a_{22} \end{bmatrix}$ is obtained by multiplication by d_1 and d_2 of the rows of A (respectively — of the columns of A). In particular, $DA = AD$ only if either the matrix A is diagonal too, or D is a **scalar** matrix (that is $d_1 = d_2$).

We see that matrix multiplication is not commutative: it can happen that $AB \neq BA$. Now — the good news: matrix multiplication is associative: $(AB)C = A(BC)$. This follows, without any further computations, from associativity of composition of abstract functions: given $h : W \to X$, $g : X \to Y$, $f : Y \to Z$, composing f with the result of composition of g and h yields exactly the same rule as composing the result of composition of f and g with h: $z = f(g(h(w)))$.

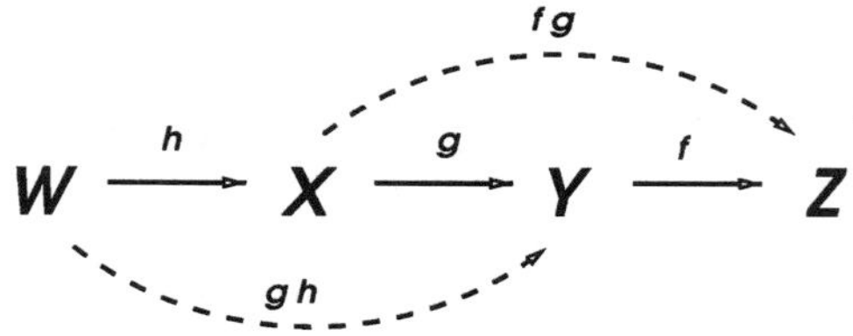

Exercises 1.3.2.

(a) Carry out the straightforward computation which yields the component-wise formulas for composition of linear transformations.

(b) Work out the details of Example (b).

(c) Show that $T_\theta R_0 = R_\theta$ and $R_0 T_\theta = R_{-\theta}$ by a matrix computation and by a geometrical argument.

(d) Find the point on the coordinate plane which is obtained from the point $(x_1, x_2) = (1, 2)$ by clockwise rotation about the origin through $\pi/3$ followed by the reflection about the line $x_1 = 0$.

(e) Compute the matrix product $T_\phi T_\psi$ and derive the addition formulas for $\cos(\phi + \psi)$ and $\sin(\phi + \psi)$.

(f) Show that $R_\phi R_\psi = T_{\phi - \psi}$.

1.3.3. Inverses. Two functions, $g : X \to Y$ and $f : Y \to X$, are called inverse to each other if $f(g(x)) = x$ and $g(f(y)) = y$ for any x and y. In other words, each of them "undoes" what the other one does. For instance, the rotation through $-\theta$ is inverse to the rotation through θ, a reflection is its own inverse.

$$X \underset{f}{\overset{g}{\rightleftarrows}} Y$$

If the inverse to a linear transformation $\mathbf{x}' = A\mathbf{x}$ exists, then A is called invertible, and the inverse transformation is denoted $\mathbf{x} = A^{-1}\mathbf{x}'$ and is also linear (why?) By the very definition the relationship between A and A^{-1} reads: $AA^{-1}\mathbf{x} = \mathbf{x} = A^{-1}A\mathbf{x}$ for any vector $\mathbf{x}$. The identity linear transformation $\mathbf{x}' = \mathbf{x}$ has the same matrix in any basis which is called the identity matrix:

$$I = \begin{bmatrix} 1 & 0 \\ 0 & 1 \end{bmatrix}.$$

Thus, in matrix terms we have $AA^{-1} = I = A^{-1}A$. It is easy to check the following explicit formula for inverse 2×2-matrices:

$$A = \begin{bmatrix} a & b \\ c & d \end{bmatrix}, \quad A^{-1} = \frac{1}{ad - bc} \begin{bmatrix} d & -b \\ -c & a \end{bmatrix}.$$

Thus, the matrix A is invertible if $ad - bc \neq 0$. If $ad - bc = 0$ then A is not invertible. Indeed, according to Exercise 1.1.3(c) if $ad = bc$ then the vectors $A\mathbf{e}_1$ and $A\mathbf{e}_2$ are proportional, while the vectors $\mathbf{e}_1, \mathbf{e}_2$ are not! Since linear transformations send proportional vectors to proportional they cannot undo the "damage" done by A.

Remark. The concept of invertible linear transformations is very similar to that of changes of coordinates. Indeed, let

$$\begin{aligned} x_1 &= ax_1' + bx_2' \\ x_2 &= cx_1' + dx_2' \end{aligned}$$

be the expression of coordinates x_1, x_2 in the basis $\mathbf{e}_1, \mathbf{e}_2$ as linear functions of new coordinates x_1', x_2' in a basis $\mathbf{f}_1, \mathbf{f}_2$. The matrix $\begin{bmatrix} a & b \\ c & d \end{bmatrix}$ here is called the transition matrix from the old coordinate system to the new one. The vectors $\mathbf{f}_1, \mathbf{f}_2$ have coordinates $\begin{bmatrix} x_1' \\ x_2' \end{bmatrix} = \begin{bmatrix} 1 \\ 0 \end{bmatrix}$ and $\begin{bmatrix} 0 \\ 1 \end{bmatrix}$ in the basis $\mathbf{f}_1, \mathbf{f}_2$. We see therefore that the columns $\begin{bmatrix} a \\ c \end{bmatrix}, \begin{bmatrix} b \\ d \end{bmatrix}$ of the transition matrix represent the vectors $\mathbf{f}_1, \mathbf{f}_2$ in the old basis $\mathbf{e}_1, \mathbf{e}_2$.[1] Respectively, the new coordinates x_1', x_2' of a vector $\mathbf{x}$ are expressed via the old coordinates x_1, x_2 by means of the inverse transition matrix.

Here we are talking about the same vectors expressed by different coordinates in different coordinate systems. On the other hand, we can use the same matrix $\begin{bmatrix} a & b \\ c & d \end{bmatrix}$ in order to define a linear transformation which transforms the vector with coordinates x_1, x_2 to a new vector with coordinates $ax_1 + bx_2,\ cx_1 + dx_2$ in *the same* coordinate system.

We illustrate the use of both languages in the following example.

---

[1]In particular, $\mathbf{f}_1, \mathbf{f}_2$ form a basis (and therefore the formulas $x_1 = ax_1' + bx_2'$, $x_2 = cx_1' + dx_2'$ define a change of coordinates) if and only if $ad - bc \neq 0$.

Examples. (a) According to the Orthogonal Diagonalization Theorem an ellipse with the equation $ax_1^2 + 2bx_1x_2 + cx_2^2 = 1$ is given by one of the normal equations $X_1^2/a_1^2 + X_2^2/a_2^2 = 1$ in a rotated coordinate system. We could rephrase it this way: any ellipse centered at the origin is obtained from a normal one by a rotation.

(b) *Any invertible linear transformation A on the plane can be represented as the composition ST: an orthogonal transformation T followed by stretching S in the directions of two perpendicular axes* (with possibly different coefficients a_1, a_2). Indeed, A transforms the unit circle to an ellipse which in its turn can obtained from the unit circle by stretching S in the directions of the principal axes. Thus, the composition $T = S^{-1}A$ preserves the unit circle and hence — lengths of all vectors. Due to Exercise 1.1.2(c) it preserves inner products of all vectors. Thus T is an orthogonal transformation.

Exercises 1.3.3.

(a) Verify the formula for inverse matrices. Check that the formula applied to T_θ and R_θ yields $T_{-\theta}$ and respectively R_θ.

(b) Find a non-invertible matrix (such matrices are called singular) whose all entries are non-zero. Find the range of the linear transformation defined by your matrix.

(c) For invertible A and B, prove that $(AB)^{-1} = B^{-1}A^{-1}$.

(d) For any integer n and invertible A put $A^n = A...A$ (n times) if n is positive, $A^n = A^{-1}...A^{-1}$ ($|n|$ times) if n is negative and $A^0 = I$. Prove that for any m, n we have $A^mA^n = A^{m+n}$.

(e) Compute the powers A^n for those n for which the power is defined, if A is diagonal,

$$A = \begin{bmatrix} 0 & 1 \\ 0 & 0 \end{bmatrix}, \begin{bmatrix} 1 & 0 \\ 0 & 0 \end{bmatrix}, \begin{bmatrix} 1 & 1 \\ 0 & 1 \end{bmatrix}.$$

(f) Rotate the normal ellipse with semiaxes $2, 1$ through $\pi/4$ counter-clockwise about the origin.

(g) Prove that any invertible linear transformation can be represented as the composition TS: stretching (possibly with different coefficients) in the directions of two perpendicular lines followed by an orthogonal transformation.

1.3.4. Matrix Zoo. I. It is convenient to consider rows and columns as 1×2- and 2×1 matrices representing linear forms and, respectively, vectors. The matrix product $\mathbf{ax}$ of the row $[a_1, a_2]$ with the column $\begin{bmatrix} x_1 \\ x_2 \end{bmatrix}$ is then the "1×1-matrix" $a_1x_1 + a_2x_2$, that is a number equal to the value of the linear form $\mathbf{a}$ on the vector $\mathbf{x}$. Also, the action of a linear transformation T on the vector $\mathbf{x}$ is described in coordinates as the matrix product $\mathbf{x}' = T\mathbf{x}$ of a 2×2-matrix T with the 2×1-matrix $\mathbf{x}$.

II. Matriices and matrix operations come handy in various situations not always directly connected to linear transformations.

Linear systems.

A system of 2 linear equations in 2 unknowns can be rewritten as a single matrix equation $A\mathbf{x} = \mathbf{b}$:

$$\begin{array}{ccccc} a_{11}x_1 & + & a_{12}x_2 & = & b_1 \\ a_{21}x_1 & + & a_{22}x_2 & = & b_2 \end{array} \Leftrightarrow \begin{bmatrix} a_{11} & a_{12} \\ a_{21} & a_{22} \end{bmatrix} \begin{bmatrix} x_1 \\ x_2 \end{bmatrix} = \begin{bmatrix} b_1 \\ b_2 \end{bmatrix}.$$

Coordinate transformations.

A change of coordinates $\mathbf{x} = A\mathbf{x}'$ (where A is the transition matrix) acts on a linear form $\mathbf{ax}$ as $\mathbf{a}(A\mathbf{x}') = (\mathbf{a}A)\mathbf{x}'$ (associativity of matrix multiplication!) Thus the matrix product formula $\mathbf{a}A$ of the 1×2-matrix with the 2×2-matrix shows how the change of coordinates affects the coefficients of the linear form.

Similarly, if T is the matrix of a linear transformation $\mathbf{y} = T\mathbf{x}$, and $\mathbf{x} = A\mathbf{x}'$, $\mathbf{y} = A\mathbf{y}'$ is the transformation of both vectors to a new coordinate system, then $\mathbf{y}' = A^{-1}TA\mathbf{x}'$. The rule $T \mapsto A^{-1}TA$ shows therefore how the change of coordinates affects matrices of linear transformation. The transformations of matrices by $T \mapsto A^{-1}TA$ are called **similarity transformations**, and the matrices T and $A^{-1}TA$ are called **similar**.

Quadratic forms.

A quadratic form $ax_1^2 + 2bx_1x_2 + cx_2^2$ can be written in terms of the inner product and the matrix product as $\langle \mathbf{x}, Q\mathbf{x}\rangle$ where Q is the square matrix $\begin{bmatrix} a & b \\ b & c \end{bmatrix}$ **symmetric** about the principal diagonal.

More generally, let A be a linear transformation. The function of two vectors

$$\langle \mathbf{u}, A\mathbf{v}\rangle = a_{11}u_1v_1 + a_{12}u_1v_2 + a_{21}u_2v_1 + a_{22}u_2v_2$$

is called a **bilinear form** for it has the same bilinearity property as the inner product. Indeed, due to linearity of A and bilinearity of the inner product, we have

$$\langle \mathbf{u}, A(\alpha\mathbf{v} + \beta\mathbf{w})\rangle = \langle \mathbf{u}, \alpha A\mathbf{v} + \beta A\mathbf{w}\rangle = \alpha\langle \mathbf{u}, A\mathbf{v}\rangle + \beta\langle \mathbf{u}, A\mathbf{w}\rangle,$$

and the other identity is even easier (check it!)

Interchanging the roles of $\mathbf{u}$ and $\mathbf{v}$ we conclude that

$$\langle \mathbf{v}, A\mathbf{u}\rangle = \langle A\mathbf{u}, \mathbf{v}\rangle = \langle \mathbf{u}, A^t\mathbf{v}\rangle, \text{ where } A^t = \begin{bmatrix} a_{11} & a_{21} \\ a_{12} & a_{22} \end{bmatrix}.$$

The matrix A^t is called **transposed** to A and is obtained from A by reflection about the diagonal. We see that the bilinear form is symmetric if and only if the matrix A is symmetric: $A^t = A$.

III. *Several formulas involving transposition.*

(a) $(AB)^t = B^tA^t$. Indeed,

$$\langle (AB)^t\mathbf{u}, \mathbf{v}\rangle = \langle \mathbf{u}, AB\mathbf{v}\rangle = \langle A^t\mathbf{u}, B\mathbf{v}\rangle = \langle B^tA^t\mathbf{u}, \mathbf{v}\rangle$$

for any $\mathbf{u}, \mathbf{v}$. This is possible only if $(AB)^t$ and B^tA^t is the same matrix. If this (correct!) proof does not convince you, check the formula by a direct computation.

(b) The definition of an orthogonal transformation U means that $\langle \mathbf{x}, \mathbf{y}\rangle = \langle U\mathbf{x}, U\mathbf{y}\rangle = \langle \mathbf{x}, U^tU\mathbf{y}\rangle$ for any $\mathbf{x}, \mathbf{y}$. Thus matrices of orthogonal transformations in an orthonormal basis are characterized by the property $U^tU = I$ or, equivalently, $U^{-1} = U^t$. Matrices with this property are called **orthogonal matrices**.

(c) The change of coordinates $\mathbf{x} = A\mathbf{x}'$ transforms the quadratic form $\langle \mathbf{x}, Q\mathbf{x}\rangle$ to $\langle A\mathbf{x}', QA\mathbf{x}'\rangle = \langle \mathbf{x}', A^tQA\mathbf{x}'\rangle$. The transformation rule $Q \mapsto A^tQA$ shows therefore how the coefficients of quadratic forms are affected by changes of coordinates.

(d) We see from above examples that symmetric matrices and transposition occur in connection with quadratic or bilinear forms written in terms of inner products. Finally, the inner product itself can be expressed as the matrix product $\langle \mathbf{u}, \mathbf{v}\rangle = \mathbf{u}^t\mathbf{v}$ if the transposition $\mathbf{u}^t$ of the column $\mathbf{u}$ is understood as the corresponding row: $\mathbf{u}^t = [u_1, u_2]$.

IV. We give several reformulations of the Orthogonal Diagonalization Theorem (yet to be proved!) Due to III (c) it has the following matrix formulation:

The symmetric matrix Q of a quadratic form can be transformed into the diagonal matrix $\begin{bmatrix} A & 0 \\ 0 & C \end{bmatrix}$ *by the transformation $Q \mapsto U^tQU$ where U is a suitable orthogonal matrix.*

Notice that $U^tQU = U^{-1}QU$ since the matrix U is orthogonal. Thus symmetric matrices of quadratic forms are transformed by changes of Cartesian coordinates in the same way as matrices of linear transformations. We can rephrase the Orthogonal Diagonalization Theorem this way:

If a linear transformation has a symmetric matrix Q in an orthonormal basis, then it has a diagonal matrix in a suitable new orthonormal basis.

A diagonal matrix defines the linear transformation of stretching (with possibly different coefficients) in the directions of the coordinate axes. [2] Therefore we have one more reformulation of the Orthogonal Diagonalization Theorem:

A linear transformation defined by a symmetric matrix is a stretch in the directions of two perpendicular lines.

V. *Matrix arithmetics.*

Sums of vectors are vectors. Sums of linear functions are linear functions. Sums of quadratic forms are quadratic forms. Sums of linear transformations, defined by the formula $(A + B)\mathbf{x} = A\mathbf{x} + B\mathbf{x}$, are linear transformations (check it!) Matrix expressions for these operations give rise to component-wise addition of matrices of the same format — columns with columns, rows with rows, square matrices with square matrices:

$$\begin{bmatrix} a_{11} & a_{12} \\ a_{21} & a_{22} \end{bmatrix} + \begin{bmatrix} b_{11} & b_{12} \\ b_{21} & b_{22} \end{bmatrix} = \begin{bmatrix} a_{11}+b_{11} & a_{12}+b_{12} \\ a_{21}+b_{21} & a_{22}+b_{22} \end{bmatrix}.$$

The arithmetics of matrix addition and multiplication is similar to that of numbers: the distributive laws $(A + B)C = AC + BC$, $C(A + B) = CA + CB$ hold true (whenever formats of the matrices allow to form the expressions). What is not true in general, is that $AB = BA$. For instance, $(A-B)(A+B) = A^2+AB-BA-B^2$, but it is equal to the usual $A^2 - B^2$ only if the matrices A and B commute.

Exercises. 1.3.4.

(a) Rotate the line $x_1 + 2x_2 = 3$ through $\pi/4$ counter-clockwise about the origin.

(b) Find the matrix of the reflection R_0 in the rotated basis $\mathbf{f}_1 = T_\theta\mathbf{e}_1$, $\mathbf{f}_2 = T_\theta\mathbf{e}_2$.

(c) Which of the following are bilinear forms? symmetric bilinear forms?

$$(u_1 + u_2)(v_1 - v_2), \quad (u_1 + v_1)(u_2 + v_2), \quad (u_1 + u_2)(v_1 + v_2)$$

(d) Prove that $(A^t)^t = A$.

(e) Compress the ellipse $5x_1^2+6x_1x_2+5x_2^2 = 2$ two times in the direction of the line $x_1+x_2 = 0$.

(f) Check that $T_\theta^t = T_\theta^{-1}$, $R_\theta^t = R_\theta^{-1}$.

(g) Which diagonal matrices are similar to each other?

(h) Represent the linear transformations defined by the following symmetric matrices as stretching in the directions of two perpendicular lines:

$$\begin{bmatrix} 1 & 0 \\ 0 & 0 \end{bmatrix}, \begin{bmatrix} 1 & 0 \\ 0 & -1 \end{bmatrix}, \begin{bmatrix} 0 & 1 \\ 1 & 0 \end{bmatrix}, \begin{bmatrix} 1 & 1 \\ 1 & 1 \end{bmatrix}.$$

(i) Let $P(x)$ be a polynomial in one variable. For similar matrices A and B, prove that $P(A)$ and $P(B)$ are also similar.

---

[2] Of course, this description is literally true only if the stretching coefficients are greater than 1. "Stretching k times" involves also a flip when k is negative, is actually shrinking if $|k| < 1$, and for $k = 0$ makes the plane collapse to a line. We use the word *stretch* in this generalized sense.

1.4. Complex numbers

The quadratic equation $x^2 - 1 = 0$ in one unknown has two solutions $x = \pm 1$. The equation $x^2+1 = 0$ has no solutions at all. For the sake of justice one introduces a new number i, the imaginary unit, such that $i^2 = -1$, and thus $x = \pm i$ become two solutions to the equation.

1.4.1. Definitions and geometrical interpretations. Complex numbers are defined as ordered pairs of real numbers written in the form $z = a + bi$. The real numbers a and b are called the real part and imaginary part of the complex number z and denoted $a = \operatorname{Re} z$ and $b = \operatorname{Im} z$. The sum of two complex numbers z and $w = c + di$ is defined by $z + w = (a + c) + (b + d)i$ while the definition of the product is to comply with the relation $i^2 = -1$:

$$zw = ac + bdi^2 + adi + bci = (ac - bd) + (ad + bc)i.$$

Operations of addition and multiplication of complex numbers enjoy the same properties as those of real numbers. In particular, the product is commutative and associative.

The complex number $\bar{z} = a - bi$ is called complex conjugate to $z = a + bi$. The formula $\overline{z + w} = \bar{z} + \bar{w}$ is obvious, and $\overline{zw} = \bar{z}\bar{w}$ is due to the fact that $\bar{i} = -i$ has exactly the same property as i: $(-i)^2 = -1$.

The product $z\bar{z} = a^2 + b^2$ (check this formula!) is real and positive unless $z = 0 + 0i = 0$. This shows that

$$\frac{1}{z} = \frac{\bar{z}}{z\bar{z}} = \frac{a}{a^2 + b^2} - \frac{b}{a^2 + b^2} i,$$

and hence the division by z is well-defined for any non-zero complex number z.

The non-negative real number $|z| = \sqrt{z\bar{z}} = \sqrt{a^2 + b^2}$ is called the absolute value of z. The absolute value function has the same multiplicative property as in the case of real numbers: $|zw| = \sqrt{zw\overline{zw}} = \sqrt{z\bar{z}w\bar{w}} = |z| \cdot |w|$. It actually coincides with the absolute value of real numbers when applied to complex numbers with zero imaginary part: $|a + 0i| = |a|$.

To a complex number $z = a+bi$, we can associate the radius-vector $\mathbf{z} = a\mathbf{e}_1+b\mathbf{e}_2$ on the coordinate plane. The unit coordinate vectors $\mathbf{e}_1$ and $\mathbf{e}_2$ represent therefore the complex numbers 1 and i. The coordinate axes are called respectively the real and imaginary axes of the plane. Addition of complex numbers coincides with the operation of vector sum.

The absolute value function has the geometrical meaning of the distance to the origin: $|z| = \langle \mathbf{z}, \mathbf{z} \rangle^{1/2}$, while $z\bar{z}$ is the inner square. In particular, the triangle inequality $|z + w| \le |z| + |w|$ holds true. Complex numbers of unit absolute value $|z| = 1$ form the unit circle centered at the origin.

The operation of complex conjugation acts on the vectors $\mathbf{z}$ as the reflection about the real axis.

In order to describe a geometrical meaning of complex multiplication, let us write the vector representing a non-zero complex number z in the polar (or trigonometric) form $z = ru$ where $r = |z|$ is a positive real number, and $u = z/|z| = \cos\theta + i \sin\theta$ has the absolute value 1. Here $\theta = \arg z$, called the argument of the complex number z, is the angle that the vector $\mathbf{z}$ makes with the positive direction of the real axis.

Consider the linear transformation on the plane defined as multiplication of all complex numbers by a given complex number z. It is the composition of the multiplication by r and by u. The geometrical meaning of multiplication by r is clear: it makes all vectors r times longer. The multiplication by u is described by the following formulas

$$\begin{array}{l}\mathrm{Re}\,[(\cos\theta+i\sin\theta)(x_1+ix_2)]=(\cos\theta)x_1-(\sin\theta)x_2\\ \mathrm{Im}\,[(\cos\theta+i\sin\theta)(x_1+ix_2)]=(\sin\theta)x_1+(\cos\theta)x_2\end{array}.$$

This is a linear transformation with the matrix $\begin{bmatrix}\cos\theta & -\sin\theta\\ \sin\theta & \cos\theta\end{bmatrix}$. The multiplication by u is therefore the rotation through the angle θ. Thus the multiplication by z is the composition of the dilation by $|z|$ and rotation through $\arg z$.

In other words, the product operation of complex numbers sums their arguments and multiplies absolute values:

$$|zw|=|z|\cdot|w|,\quad \arg zw=\arg z+\arg w \text{ modulo } 2\pi.$$

1.4.2. The exponential function. Consider the series

$$1+z+\frac{z^2}{2}+\frac{z^3}{6}+\ldots+\frac{z^n}{n!}+\ldots$$

Applying the ratio test for convergence of infinite series,

$$|\frac{z^n(n-1)!}{n!z^{n-1}}|=\frac{|z|}{n}\to 0<1 \text{ as } n\to\infty,$$

we conclude that the series converges absolutely for any complex number z. The sum of the series is a complex number denoted $\exp z$, and the rule $z\mapsto\exp z$ defines the exponential function of the complex variable z.

The exponential function transforms sums to products:

$$\exp(z+w)=(\exp z)(\exp w) \text{ for any complex } z \text{ and } w.$$

Indeed, due to the binomial formula, we have

$$(z+w)^n=\sum_{k=0}^{n}\binom{n}{k}z^k w^{n-k}=n!\sum_{k+l=n}\frac{z^k}{k!}\frac{w^l}{l!}.$$

Rearranging the sum over all n as a double sum over k and l we get

$$\sum_{n=0}^{\infty}\frac{(z+w)^n}{n!}=\sum_{k=0}^{\infty}\sum_{l=0}^{\infty}\frac{z^k}{k!}\frac{w^l}{l!}=(\sum_{k=0}^{\infty}\frac{z^k}{k!})(\sum_{l=0}^{\infty}\frac{w^l}{l!}).$$

The exponentials of complex conjugated numbers are conjugated:

$$\exp\bar z=\sum\frac{\bar z^n}{n!}=\overline{\sum\frac{z^n}{n!}}=\overline{\exp z}.$$

In particular, on the real axis the exponential function is real and coincides with the usual real exponential function $\exp x=e^x$ where $e=1+1/2+1/6+\ldots+1/n!+\ldots=\exp(1)$. Extending this notation to complex numbers we can rewrite the above properties of $e^z=\exp z$ as $e^{z+w}=e^z e^w$, $e^{\bar z}=\overline{e^z}$.

On the imaginary axis, $w=e^{iy}$ satisfies $w\bar w=e^0=1$ and hence $|e^{iy}|=1$. The way the imaginary axis is mapped by the exponential function to the unit circle is described by the following Euler's formula:

$$e^{i\theta}=\cos\theta+i\sin\theta.$$

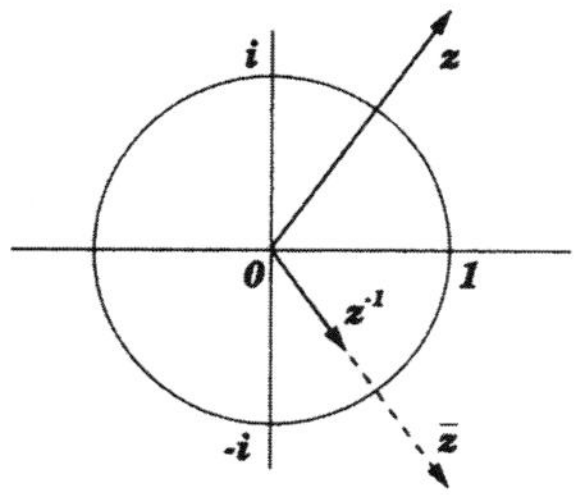

Exercises 1.4.1.

(a) Compute $(1+i)/(3-2i)$, $(\cos\pi/3+i\sin\pi/3)^{-1}$.

(b) Show that z^{-1} is a real proportional to $\bar{z}$ and find the proportionality coefficient.

(c) Find all z satisfying $|z-1|=|z-2|=1$.

(d) Sketch the solution set to the following system of inequalities: $|z+1|\leq 1$, $|z|\leq 1$, $\mathrm{Re}(iz)\leq 0$.

(e) Compute $(\frac{\sqrt{3}+i}{2})^{100}$.

(f) Prove that the linear transformation defined by the matrix $\begin{bmatrix} a & -b \\ b & a \end{bmatrix}$ is the composition of multiplication by $\sqrt{a^2+b^2}$ and a rotation.

(g) Let $z_1,...,z_5$ form a regular pentagon inscribed into the unit circle $|z|=1$. Prove that $z_1+...+z_5=0$.

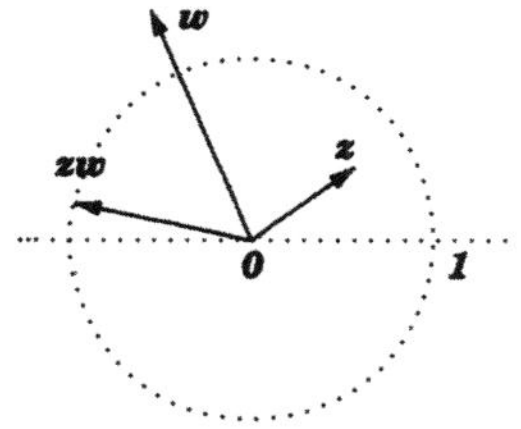

Exercises 1.4.2.

(a) Prove the "Fundamental Formula of Mathematics": $e^{\pi i}+1=0$.

(b) Represent $1-i$ and $1-\sqrt{3}i$ in the polar form $re^{i\theta}$.

(c) Show that $\cos\theta=(e^{i\theta}+e^{-i\theta})/2$ and $\sin\theta=(e^{i\theta}-e^{-i\theta})/2i$.

(d) Compute the real and imaginary part of the product $e^{i\phi}e^{i\psi}$ using the Euler formula and deduce the addition formulas for $\cos(\phi+\psi)$ and $\sin(\phi+\psi)$.

(e) Express $\mathrm{Re}\, e^{3i\theta}$, $\mathrm{Im}\, e^{3i\theta}$ in terms of $\mathrm{Re}\, e^{i\theta}$ and $\mathrm{Im}\, e^{i\theta}$ and deduce the triple argument formulas for $\cos 3\theta$ and $\sin 3\theta$.

(f) Prove the binomial formula:

$$(z+w)^n=\sum_{k=0}^{n}\binom{n}{k}z^k w^{n-k},$$

where $\binom{n}{k}=n!/k!(n-k)!$.

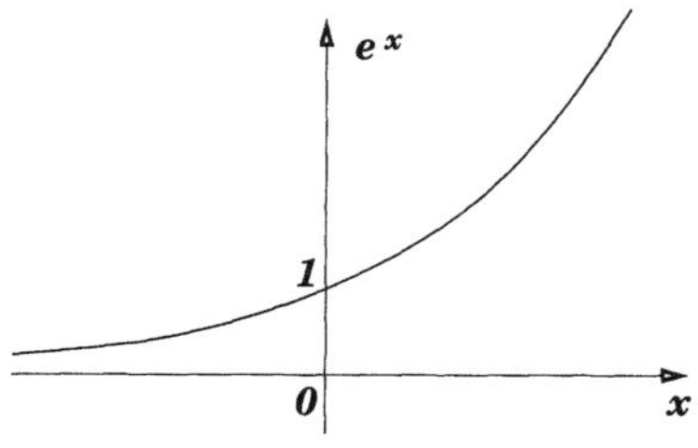

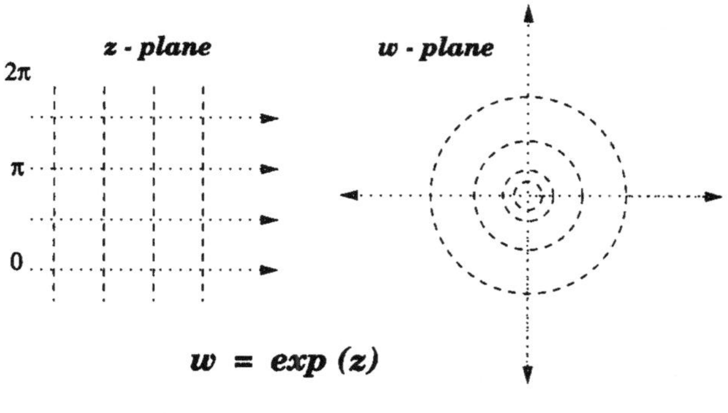

It is proved by comparison of $e^{i\theta} = \sum (i\theta)^n/n!$ with Taylor series for $\cos\theta$ and $\sin\theta$:

$$\begin{aligned} \operatorname{Re} e^{i\theta} &= 1 - \tfrac{\theta^2}{2} + \tfrac{\theta^4}{24} - ... = \sum (-1)^k \tfrac{\theta^{2k}}{(2k)!} = \cos\theta \\ \operatorname{Im} e^{i\theta} &= \theta - \tfrac{\theta^3}{6} + \tfrac{\theta^5}{120} - ... = \sum (-1)^k \tfrac{\theta^{2k+1}}{(2k+1)!} = \sin\theta \end{aligned} .$$

Thus $\theta \mapsto e^{i\theta}$ is the usual parameterization of the unit circle by the angular coordinate θ. In particular, $e^{2\pi i} = 1$ and therefore the exponential function is $2\pi i$-periodic: $e^{z+2\pi i} = e^z e^{2\pi i} = e^z$. Using Euler's formula we can rewrite the polar form of a non-zero complex number w as

$$w = |w| e^{i \arg w}.$$

1.4.3. The Fundamental Theorem of Algebra. A quadratic polynomial $z^2 + pz + q$ has two roots

$$z_{\pm} = \frac{-p \pm \sqrt{p^2 - 4q}}{2}$$

regardless of the sign of the discriminant $p^2 - 4q$, if we allow the roots to be complex and take in account multiplicity. Namely, if $p^2 - 4q = 0$, $z^2 + pz + q = (z + p/2)^2$ and therefore the single root $z = -p/2$ has multiplicity two. If $p^2 - 4q < 0$ the roots are complex conjugated with $\operatorname{Re} z_{\pm} = -p/2$, $\operatorname{Im} z_{\pm} = \pm\sqrt{|p^2 - 4q|}/2$. The Fundamental Theorem of Algebra shows that not only the justice has been restored, but that any degree n polynomial has n complex roots, possibly — multiple.

Theorem. *A degree n polynomial $P(z) = z^n + a_1 z^{n-1} + ... + a_{n-1} z + a_n$ with complex coefficients $a_1, ..., a_n$ factors as*

$$P(z) = (z - z_1)^{m_1} ... (z - z_r)^{m_r}.$$

Here $z_1, ..., z_r$ are complex roots of $P(z)$ of multiplicities $m_1, ..., m_r$, and $m_1 + ... + m_r = n$.

This is one of a few theorems we intend to use in this course without proof. We illustrate it with the following examples.

Examples. (a) The equation $z^2 = w$, where $w = re^{i\theta}$ is a complex number written in the polar form, has two solutions $\pm\sqrt{w} = \pm\sqrt{r} e^{i\theta/2}$. Thus the formula for roots of quadratic polynomials makes sense even if the coefficients p, q are complex.

(b) The complex numbers $1, i, -1, -i$ are the roots of the polynomial $z^4 - 1 = (z^2 - 1)(z^2 + 1) = (z - 1)(z + 1)(z - i)(z + i)$.

(c) There are n complex n-th roots of unity. Namely, if $z = re^{i\theta}$ satisfies $z^n = 1$ then $r^n e^{in\theta} = 1$ and therefore $r = 1$ and $n\theta = 2\pi k$, $k = 0, \pm 1, \pm 2, ...$ Thus

$$z = e^{2\pi i k/n} = \cos\frac{2\pi k}{n} + i \sin\frac{2\pi k}{n}, \quad k = 0, 1, 2, ..., n-1.$$

For instance, if $n = 3$, the roots are 1 and

$$e^{\pm 2\pi i/3} = \cos\frac{2\pi}{3} \pm i \sin\frac{2\pi}{3} = -\frac{1}{2} \pm i\frac{\sqrt{3}}{2}.$$

As illustrated by the previous two examples, if the coefficients $a_1, ..., a_n$ of the polynomial $P(z)$ are real numbers, that is $\bar{a}_i = a_i$, yet the roots can be non-real, but then they come in complex conjugated pairs. This follows from equality of two factorizations for $\overline{P(\bar{z})} = z^n + \bar{a}_1 z^{n-1} + ... + \bar{a}_n = P(z)$:

$$(z - \bar{z}_1)^{m_1} ... (z - \bar{z}_r)^{m_r} = (z - z_1)^{m_1} ... (z - z_r)^{m_r}.$$

These equal products can differ only by the order of the factors, and thus for each non-real root of $P(z)$ the complex conjugate is also a root and of the same multiplicity.

Expanding the product

$$(z - z_1)...(z - z_n) = z^n - (z_1 + ... + z_n)z^{n-1} + ... + (-1)^n z_1...z_n$$

we can express coefficients $a_1, ..., a_n$ of the polynomial via the roots $z_1, ..., z_n$ (here multiple roots should be repeated according to their multiplicities). In particular, the sum and the product of roots are

$$z_1 + ... + z_n = -a_1, \quad z_1...z_n = (-1)^n a_n.$$

These formulas generalize the Vieta theorem for roots of quadratic polynomials: $z_+ + z_- = -p$, $z_+ z_- = q$.

Exercises 1.4.3.

(a) Solve the quadratic equations:

$$z^2 - 6z + 5 = 0, \quad z^2 - iz + 1 = 0, \quad z^2 - 2(1+i)z + 2i = 0, \quad z^2 - 2z + \sqrt{3}i = 0.$$

(b) Solve the equations

$$z^3 + 8 = 0, \quad z^3 + i = 0, \quad z^4 + 4z^2 + 4 = 0, \quad z^4 - 2z^2 + 4 = 0, \quad z^6 + 1 = 0.$$

(c) Prove that for any $n > 1$ the sum of all n-th roots of unity equals zero.

(d) Prove that any polynomial with real coefficients factors into a product of linear and quadratic polynomials with real coefficients.

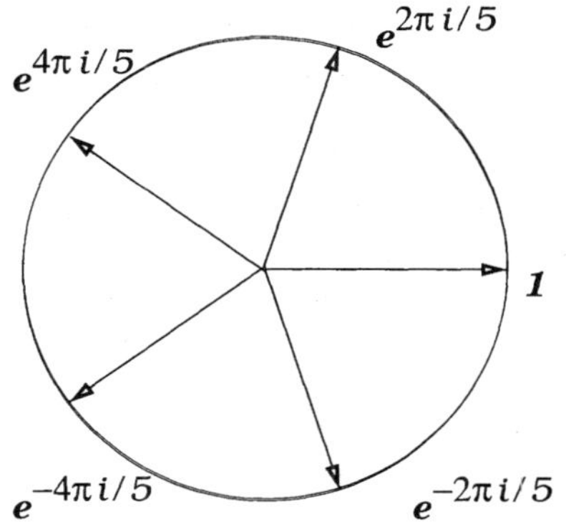

5-th roots of unity

1.5. Eigenvalues

In this section, after some preparation, we prove the Orthogonal Diagonalization Theorem and classify linear transformations on the plane up to changes of coordinates.

1.5.1. Linear systems. Consider the system of two linear equations in two unknowns:

$$\begin{aligned} a_{11}z_1 + a_{12}z_2 &= b_1 \\ a_{21}z_1 + a_{22}z_2 &= b_2 \end{aligned}.$$

In order to solve it, let us multiply the 1-st equation by a_{22} and subtract the 2-nd equation multiplied by a_{12}. We get

$$(a_{11}a_{22} - a_{12}a_{21})\, z_1 = b_1a_{22} - b_2a_{12}.$$

Similarly, subtracting the 1-st equation multiplied by a_{21} from the 2-nd equation multiplied by a_{11} we find

$$(a_{11}a_{22} - a_{12}a_{21})\, z_2 = a_{11}b_2 - a_{21}b_1.$$

We conclude that *if* $a_{11}a_{22} - a_{12}a_{21} \neq 0$, *the system has a unique solution*

$$z_1 = \frac{b_1a_{22} - b_2a_{12}}{a_{11}a_{22} - a_{12}a_{21}}, \quad z_2 = \frac{a_{11}b_2 - a_{21}b_1}{a_{11}a_{22} - a_{12}a_{21}}.$$

In the next section we will point out a better way to comprehend these formulas than just by memorizing the order of subscripts. Meanwhile our point is that the computation and the final formulas make sense even if the coefficients a_{ij}, b_j are complex numbers. In this case the solution (z_1, z_2) also consists of complex numbers.

What happens when $a_{11}a_{22} - a_{12}a_{21} = 0$? If it is the case, then the linear functions $a_{11}z_1 + a_{12}z_2$ and $a_{21}z_1 + a_{22}z_2$ on the left hand side are proportional. Indeed, the 2-nd function is proportional to the 1-st one with the coefficient $k = a_{21}/a_{11} = a_{22}/a_{12}$ (unless $a_{11} = a_{12} = 0$ in which case the 1-st function is proportional to the 2-nd one with the coefficient 0). [3] The coefficient k is a complex number if coefficients of the system are complex.

The answer to the question depends now on the right hand side: the system can be inconsistent (if b_1 and b_2 are not in the same proportion) or have infinitely many solutions. Leaving the analysis of all possible situations to the reader, we formulate the answer in the special case of systems with zero right hand sides $b_1 = b_2 = 0$. Such systems are called homogeneous and are always consistent since they have the trivial solution $(z_1, z_2) = (0, 0)$.

If $a_{11}a_{22} - a_{12}a_{21} \neq 0$, *the homogeneous system has only the the trivial solution. If* $a_{11}a_{22} - a_{12}a_{21} = 0$, *then the homogeneous system has a non-trivial solution, and any other solution is proportional to it, unless all the 4 coefficients are equal to zero in which case any* (z_1, z_2) *is a solution.* The proportionality coefficient here is generally speaking a complex number too.

Systems of linear equations will arise throughout the course mainly as a tool for solving various problems of linear algebra and carrying out coordinate computations.

[3] Notice that if only one of the coefficients a_{11}, a_{12} vanishes, say $a_{12} = 0$, but $a_{11} \neq 0$, then the equality $a_{11}a_{22} = a_{12}a_{21} = 0$ shows that $a_{22} = 0$ and thus the 2-nd equation is still proportional to the 1-st one with the coefficient $k = a_{21}/a_{11}$.

Examples. (a) We have mentioned in Exercises 1.2.1 that solving a system of two linear equations in two unknowns is interpreted geometrically as finding common points of two lines on the plane.

(b) Another interpretation: representing a vector $\mathbf{b} = \begin{bmatrix} b_1 \\ b_2 \end{bmatrix}$ as a linear combination $\mathbf{b} = z_1\mathbf{a}_1 + z_2\mathbf{a}_2$ of the vectors $\mathbf{a}_1 = \begin{bmatrix} a_{11} \\ a_{21} \end{bmatrix}$ and $\mathbf{a}_2 = \begin{bmatrix} a_{12} \\ a_{22} \end{bmatrix}$ is equivalent to finding a solution to the above system.

(c) *Problem: Given a linear transformation A on the plane, find all non-zero vectors $\mathbf{v} \neq \mathbf{0}$ which are transformed to vectors proportional to them: $A\mathbf{v} = \lambda\mathbf{v}$.* Such vectors are called **eigenvectors of** A and the proportionality coefficient λ is called the corresponding **eigenvalue**. In the matrix form, we are looking for a number λ such that the homogeneous system $(\lambda I - A)\mathbf{v} = \mathbf{0}$ of two linear equations in two unknowns — the coordinates of $\mathbf{v}$ — has a non-trivial solution. As we know from the above analysis, such a solution exists only if

$$(\lambda - a_{11})(\lambda - a_{22}) - a_{12}a_{21} = 0.$$

This equation in λ is called the **characteristic equation** of the matrix A. Thus we first have to find the roots of the quadratic characteristic polynomial and then for each root λ — to solve the homogeneous linear system $(\lambda I - A)\mathbf{v} = \mathbf{0}$.

(d) Let us carry out this plan for the reflection matrix R_θ. The characteristic polynomial in question

$$(\lambda - \cos\theta)(\lambda + \cos\theta) - \sin^2\theta = \lambda^2 - 1$$

has the roots $\lambda_\pm = \pm 1$. For $\lambda = 1$ the homogeneous linear system reads

$$\begin{array}{rcrcl} (1-\cos\theta)v_1 & - & (\sin\theta)v_2 & = & 0 \\ -(\sin\theta)v_1 & + & (1+\cos\theta)v_2 & = & 0 \end{array}.$$

Since $1-\cos\theta = 2\sin^2\frac{\theta}{2}$, $1+\cos\theta = 2\cos^2\frac{\theta}{2}$, $\sin\theta = 2\sin\frac{\theta}{2}\cos\frac{\theta}{2}$, both equations are actually proportional to $(\sin\frac{\theta}{2})v_1 - (\cos\frac{\theta}{2})v_2 = 0$. We find $v_2 = (\tan\frac{\theta}{2})v_1$ where v_1 is arbitrary. Thus $(v_1, v_2) = (1, \tan\frac{\theta}{2})$ is a non-trivial solution, and any other solution is proportional to it. The result is not surprising: the symmetry line of the reflection R_θ has the slope $\tan\frac{\theta}{2}$ and consists of vectors $\mathbf{v}$ satisfying $R_\theta\mathbf{v} = \mathbf{v}$, that is — of eigenvectors with the eigenvalue $\lambda = 1$. The perpendicular line with the slope $\tan(\frac{\pi}{2} + \frac{\theta}{2})$ consists of the eigenvectors corresponding to $\lambda = -1$.

(e) In the case of the rotation T_θ the characteristic polynomial is

$$(\lambda - \cos\theta)^2 + \sin^2\theta = \lambda^2 - 2(\cos\theta)\lambda + 1.$$

The roots

$$\lambda_\pm = \cos\theta \pm \sqrt{\cos^2\theta - 1} = \cos\theta \pm i\sin\theta = e^{\pm i\theta}$$

are non-real if we assume that $\theta \neq \pi n$. Thus rotations have no eigenvectors. This is not surprising since rotations rotate directions of vectors. However we can solve the homogeneous systems $(\lambda_\pm - T_\theta)\mathbf{v} = \mathbf{0}$ for complex solutions:

$$(e^{\pm i\theta} - \cos\theta)v_1 + (\sin\theta)v_2 = 0 \text{ with } \sin\theta \neq 0$$

yields $v_1 = \pm iv_2$ since $e^{\pm i\theta} - \cos\theta = \pm i\sin\theta$. The other equation $-(\sin\theta)v_1 + (e^{\pm i\theta} - \cos\theta)v_2 = 0$ yields the same result since it is actually proportional to the first one. We conclude that all rotations T_θ have the same **complex eigenvectors**

— complex multiples of $(v_1, v_2) = (i, 1)$ and $(-i, 1)$ corresponding to the complex eigenvalues $e^{i\theta}$ and $e^{-i\theta}$ respectively.

The last example suggests that in real geometry it might be useful, or even necessary, to consider vectors with complex coordinates, matrices with complex entries, etc. Even though such complex objects may have little pictorial sense, the properties of algebraic operations with them — linear combinations of vectors, products of matrices, etc. — are exactly the same as those for real objects. The advantage of this point of view is due to the unifying power of the Fundamental Theorem of Algebra. Example: *Any square matrix A has complex eigenvectors.* Indeed, the characteristic polynomial has a complex root λ, hence the homogeneous linear system $(\lambda I - A)\mathbf{v} = \mathbf{0}$ has a non-trivial solution $\mathbf{v} \neq \mathbf{0}$, hence $A\mathbf{v} = \lambda\mathbf{v}$.

Exercises 1.5.1.
Find eigenvalues and eigenvectors of the following matrices:

$$\begin{bmatrix} \pm 1 & 0 \\ 0 & \pm 1 \end{bmatrix}, \begin{bmatrix} 0 & \pm 1 \\ \pm 1 & 0 \end{bmatrix}, \begin{bmatrix} 0 & 1 \\ 0 & 0 \end{bmatrix}, \begin{bmatrix} 1 & 0 \\ 1 & 1 \end{bmatrix}, \begin{bmatrix} 0 & 1 \\ 1 & 1 \end{bmatrix},$$

$$\begin{bmatrix} 1 & 2 \\ 2 & -2 \end{bmatrix}, \begin{bmatrix} 1 & 1 \\ 4 & -2 \end{bmatrix}, \begin{bmatrix} 1 & -1 \\ -4 & -1 \end{bmatrix}, \begin{bmatrix} 0 & 1 \\ -q & -p \end{bmatrix}, \begin{bmatrix} a & b \\ 0 & c \end{bmatrix}.$$

1.5.2. Determinants. *Definition.* The number $\det A = a_{11}a_{22} - a_{12}a_{21}$ is called the determinant of the square matrix A with the entries a_{ij}.

There following equality of numbers exhibits three styles of notation for determinants:

$$\det A = \det \begin{bmatrix} a_{11} & a_{12} \\ a_{21} & a_{22} \end{bmatrix} = \begin{vmatrix} a_{11} & a_{12} \\ a_{21} & a_{22} \end{vmatrix}.$$

We have encountered determinants several times in various situations.

Examples. (a) Proportionality criterion for two vectors $\begin{bmatrix} a \\ c \end{bmatrix}$ and $\begin{bmatrix} b \\ d \end{bmatrix}$ or for two linear forms $ax_1 + bx_2$ and $cx_1 + dx_2$ reads $ad - bc = 0$.

(b) A square matrix A is invertible if and only if $\det A \neq 0$.

(c) Typical level curves of a quadratic form $ax_1^2 + 2bx_1x_2 + cx_2^2$ are ellipses if the determinant $\begin{vmatrix} a & b \\ b & c \end{vmatrix} = ac - b^2$ is positive and are hyperbolas if it is negative (see Exercise 1.2.3(d)).

(d) The formulas in 1.5.1 for solutions to linear systems can be rewritten in terms of determinants as

$$z_1 = \begin{vmatrix} b_1 & a_{12} \\ b_2 & a_{22} \end{vmatrix} \Big/ \begin{vmatrix} a_{11} & a_{12} \\ a_{21} & a_{22} \end{vmatrix}, \quad z_2 = \begin{vmatrix} a_{11} & b_1 \\ a_{21} & b_2 \end{vmatrix} \Big/ \begin{vmatrix} a_{11} & a_{12} \\ a_{21} & a_{22} \end{vmatrix}.$$

(e) The characteristic polynomial of a square matrix A has been defined in 1.5.1 as

$$\det(\lambda I - A) = \begin{vmatrix} \lambda - a_{11} & -a_{12} \\ -a_{21} & \lambda - a_{22} \end{vmatrix} = \lambda^2 - (\operatorname{tr} A)\lambda + \det A,$$

where the trace $\operatorname{tr} A = a_{11} + a_{22}$ of A is the notation for the sum of the diagonal entries.

(f) According to Exercise 1.1.3(c) the determinant $\begin{vmatrix} u_1 & v_1 \\ u_2 & v_2 \end{vmatrix}$ equals the signed area of the parallelogram with the vertices $\mathbf{0}, \mathbf{u}, \mathbf{u} + \mathbf{v}, \mathbf{v}$.

The most important property of determinants is their multiplicativity: *The determinant of the product of two square matrices equals the product of their determinants:* $\det AB = (\det A)(\det B)$.

Is not hard to check the property by a straightforward computation with the product of 2×2-matrices. We leave this computation as an exercise. When it is done, many important corollaries follow immediately.

Examples. (g) $\det A^{-1} = (\det A)^{-1}$. Indeed $I = AA^{-1}$ implies $1 = \det I = (\det A)(\det A^{-1})$. In particular invertible matrices must have non-zero determinants. The converse statement that matrices with non-zero determinants are invertible is assured by the explicit formula for A^{-1} (given in 1.3.3) since the formula makes sense whenever $\det A \neq 0$.

(h) Similar matrices have the same characteristic polynomials. Indeed,

$$\begin{aligned} &\det(\lambda I - A^{-1}TA) = \det(A^{-1}(\lambda I - T)A) = (\det A^{-1})\det(\lambda I - T)(\det A) \\ &= \det(\lambda I - T) \end{aligned}.$$

This shows that the characteristic polynomial of T is not affected by changes of coordinates and thus depends only on the linear transformation defined by T.

(i) In particular the trace and determinant of the matrix of a linear transformation do not depend on the choice of the basis (while the matrix does). For determinants, this invariance can be explained by the following coordinateless interpretation: the number $\det T$ shows how the linear transformation affects signed areas of parallelograms. Indeed, if columns of V represent two vectors $\mathbf{u}, \mathbf{v}$, then columns of TV represent the vectors $T\mathbf{u}, T\mathbf{v}$. The signed area of the transformed parallelogram is therefore proportional to the signed area of the original parallelogram with the coefficient $\det T$: $\det TV = (\det T)(\det V)$.

(j) $\det T_\theta = 1$, $\det R_\theta = -1$. The fact that orthogonal transformations U have $\det U = \pm 1$ means geometrically that rotations and reflections preserve (unsigned) areas. It also follows algebraically from $U^tU = I$ since transposed matrices obviously have equal determinants and therefore $(\det U)^2 = 1$.

(k) A change of coordinates $\mathbf{x} = A\mathbf{x}'$ in a quadratic form transforms the symmetric matrix Q of the quadratic form to A^tQA and therefore does not change the *sign* of the determinant: $\det A^tQA = (\det A^t)(\det Q)(\det A) = (\det Q)(\det A)^2$. This explains why the sign of $\det Q$ distinguishes elliptic, hyperbolic and degenerate cases of level curves: according to the Corollary in 1.2.3 any quadratic form can be transformed to one the normal forms (elliptic: $\pm(x_1^2 + x_2^2)$, hyperbolic: $x_1^2 - x_2^2$, or degenerate: x_1^2, $-x_2^2$, 0) whose matrices have respectively positive, negative and zero determinants.

(l) Orthogonal changes of coordinates $\mathbf{x} = U\mathbf{x}'$ affect the symmetric matrix Q of a quadratic form by similarity transformations $U^tQU = U^{-1}QU$ and hence preserve the characteristic polynomial of Q. According to Orthogonal Diagonalization Theorem the quadratic form can be transformed by such changes of coordinates to one of the normal forms $AX_1^2 + CX_2^2$ with the characteristic polynomial $(\lambda - A)(\lambda - C)$. Thus the coefficients A, C in the normal form are roots of the characteristic polynomial

$$\begin{vmatrix} \lambda - a & -b \\ -b & \lambda - c \end{vmatrix} = \lambda^2 - (a+c)\lambda + (ac - b^2).$$

We complete this section with a proof of Orthogonal Diagonalization Theorem. It has several reformulations (see Principal Axes Theorem in 1.2.2, Example 1.3.3(a) and Section 1.3.4(IV)) equivalent to the following one:

For any symmetric matrix Q there exists an orthogonal matrix U such that $U^{-1}QU$ is diagonal.

Proof. The discriminant of the characteristic polynomial of Q equals

$$(a+c)^2 - 4(ac-b^2) = a^2 + 2ac + c^2 - 4ac + 4b^2 = (a-c)^2 + 4b^2$$

and is non-negative. This means that the characteristic polynomial of a symmetric matrix has a real root λ_1 and hence — a real eigenvector, a non-trivial solution to the homogeneous linear system $Q\mathbf{v}_1 = \lambda_1 \mathbf{v}_1$. Let $\mathbf{v}_2$ be a non-zero vector orthogonal to $\mathbf{v}_1$. Then

$$\langle Q\mathbf{v}_2, \mathbf{v}_1\rangle = \langle \mathbf{v}_2, Q\mathbf{v}_1\rangle = \langle \mathbf{v}_2, \lambda_1\mathbf{v}_1\rangle = \lambda_1\langle \mathbf{v}_2, \mathbf{v}_1\rangle = 0.$$

Thus $Q\mathbf{v}_2$ is also orthogonal to $\mathbf{v}_1$ and is therefore proportional to $\mathbf{v}_2$: $Q\mathbf{v}_2 = \lambda_2\mathbf{v}_2$ with some real coefficient λ_2. Normalizing the perpendicular eigenvectors $\mathbf{v}_1$ and $\mathbf{v}_2$ to the unit length we get an orthonormal basis $\mathbf{u}_1 = \mathbf{v}_1/|\mathbf{v}_1|$, $\mathbf{u}_2 = \mathbf{v}_2/|\mathbf{v}_2|$ of eigenvectors. Let U denote the matrix whose columns represent $\mathbf{u}_1$ and $\mathbf{u}_2$. It is orthogonal since

$$U^tU = \begin{bmatrix} \mathbf{u}_1^t\mathbf{u}_1 & \mathbf{u}_1^t\mathbf{u}_2 \\ \mathbf{u}_2^t\mathbf{u}_1 & \mathbf{u}_2^t\mathbf{u}_2 \end{bmatrix} = \begin{bmatrix} 1 & 0 \\ 0 & 1 \end{bmatrix}.$$

The columns of QU represent $\lambda_1\mathbf{u}_1$ and $\lambda_2\mathbf{u}_2$. Thus $QU = U\Lambda$, where $\Lambda = \begin{bmatrix} \lambda_1 & 0 \\ 0 & \lambda_2 \end{bmatrix}$ is the diagonal matrix of eigenvalues, and hence $U^{-1}QU = \Lambda$.

Exercises 1.5.2.

(a) Verify the multiplicative property of determinants.

(b) Give an example of two matrices which are not similar but have the same characteristic polynomial.

(c) Show that $\operatorname{tr}(A+B) = \operatorname{tr} A + \operatorname{tr} B$, $\operatorname{tr}(AB - BA) = 0$.

(d) Find the intersection point of the lines $3x_1 + 2x_2 = 1$ and $x_1 + 2x_2 = 3$. Find the coordinates of the intersection point in the basis $\mathbf{f}_1 = (1,3)$, $\mathbf{f}_2 = (3,1)$.

(e) Find out which of the following quadratic curves is an ellipse and which is a hyperbola and compute their semiaxes (that is the coefficients a_1, a_2 in the normal forms $x_1^2/a_1^2 \pm x_2^2/a_2^2 = 1$):

$$2x_1^2 + 3x_1x_2 + x_2^2 = 1, \quad 3x_1^2 - 3x_1x_2 + x_2^2 = 1.$$

(f) Following the procedure in the proof of the Orthogonal Diagonalization Theorem find orthogonal matrices U which diagonalize the symmetric matrices

$$\begin{bmatrix} 1 & -1 \\ -1 & 1 \end{bmatrix}, \begin{bmatrix} 0 & 1 \\ 1 & 1 \end{bmatrix}.$$

(g) Find a rotation that transforms the hyperbola $x_1^2 + 4x_1x_2 - 2x_2^2 = 1$ to one of the normal ones.

(h) Show that the following ellipses are congruent and find a rotation which transforms the first one to the other:

$$8x_1^2 - 4x_1x_2 + 5x_2^2 = 1, \quad 6x_1^2 + 2\sqrt{6}x_1x_2 + 7x_2^2 = 1.$$

(i) Prove that symmetric matrices with the same characteristic polynomial are similar.

(j) Show that the surface given by the equation $ac = b^2$ is a quadratic cone and divides the a, b, c-space into 3 regions. Sketch the surface and point out the regions corresponding to the quadratic forms $ax_1^2 + 2bx_1x_2 + cx_2^2$ which have a minimum, a maximum, a saddle. Which points in the picture correspond to the quadratic forms

$$x_1^2 + x_2^2, \quad x_1^2 - x_2^2, \quad -x_1^2 - x_2^2, \quad x_1^2, \quad -x_2^2, \quad 0\ ?$$

1.5.3. Normal forms. A linear transformation on the line is a dilation $x \mapsto \lambda x$ with some coefficient λ. In this sense a linear transformation T on the plane acts on its eigenvectors in a one-dimensional fashion: $T\mathbf{v} = \lambda\mathbf{v}$. We study eigenvectors of T in a hope to describe the action of T in one-dimensional terms.

Let λ_+, λ_- be the roots of the characteristic polynomial $\lambda^2 - (\operatorname{tr} T)\lambda + \det T$.

Case 1: *λ_+ and λ_- are real and distinct.* Then each of the homogeneous linear systems $(\lambda_\pm I - T)\mathbf{v} = \mathbf{0}$ has a non-trivial real solution $\mathbf{v}_\pm$. The non-zero eigenvectors $\mathbf{v}_+$ and $\mathbf{v}_-$ are non-proportional to one another (since $\lambda_+ \neq \lambda_-$), hence form a basis. Writing arbitrary vectors $\mathbf{x}$ as linear combination $\mathbf{x} = x_+\mathbf{v}_+ + x_-\mathbf{v}_-$ we see that $T\mathbf{x} = x_+T\mathbf{v}_+ + x_-T\mathbf{v}_- = \lambda_+x_+\mathbf{v}_+ + \lambda_-x_-\mathbf{v}_-$. Thus the matrix of T in the basis of the eigenvectors is $\begin{bmatrix} \lambda_+ & 0 \\ 0 & \lambda_- \end{bmatrix}$, and the linear transformation acts on the plane by stretching with coefficients $\lambda_\pm$ in the directions of the eigenvectors.

Case 2: *$\lambda_\pm$ is a pair of complex conjugated numbers λ and $\bar{\lambda}$.* In particular $\lambda_\pm$ are distinct. Let $\mathbf{v} \neq \mathbf{0}$ be a complex eigenvector of the matrix T: $T\mathbf{v} = \lambda\mathbf{v}$. Replacing all numbers in this equality by their complex conjugates we obtain $\overline{T}\bar{\mathbf{v}} = \bar{\lambda}\bar{\mathbf{v}}$. Since the matrix T is real we find that $\bar{\mathbf{v}}$ is the eigenvector of $\overline{T} = T$ with the eigenvalue $\bar{\lambda}$. Thus T acts on vectors with complex coordinates in the basis $\mathbf{v}, \bar{\mathbf{v}}$ of complex conjugated eigenvectors as coordinate-wise multiplication by λ and $\bar{\lambda}$.

Let us see what it means in purely real terms. Let $\mathbf{v}_\pm$ be two real vectors obtained as the real and imaginary parts of $\mathbf{v}$: $\mathbf{v}_+ = (\mathbf{v}+\bar{\mathbf{v}})/2$, $\mathbf{v}_- = (\mathbf{v}-\bar{\mathbf{v}})/2i$. The vectors $\mathbf{v}_\pm$ are non-proportional to each other since $\mathbf{v} = \mathbf{v}_+ + i\mathbf{v}_-$ and $\bar{\mathbf{v}} = \mathbf{v}_+ - i\mathbf{v}_-$ are non-proportional as complex vectors. Let $\lambda = a + ib = r(\cos\theta + i\sin\theta)$. We have

$$T\mathbf{v}_+ = \frac{1}{2}(\lambda\mathbf{v} + \bar{\lambda}\bar{\mathbf{v}}) = \frac{\lambda + \bar{\lambda}}{2}\mathbf{v}_+ + i\frac{\lambda - \bar{\lambda}}{2}\mathbf{v}_- = r\cos\theta\ \mathbf{v}_+ + r\sin\theta\ \mathbf{v}_-,$$

and similarly $T\mathbf{v}_- = -r\sin\theta\ \mathbf{v}_+ + r\cos\theta\ \mathbf{v}_-$. The matrix of T in the basis $\mathbf{v}_+, \mathbf{v}_-$ is rT_θ and thus the action of T in the corresponding coordinate system looks like the complex multiplication by λ: composition of the rotation through $\arg\lambda$ and the dilation $|\lambda|$ times.

Case 3: *multiple root λ.* The root is real (why?) If the homogeneous system $(\lambda I - T)\mathbf{v} = \mathbf{0}$ is satisfied by any vector $\mathbf{v}$ then $T = \lambda I$ is the scalar matrix, and T acts as the multiplication by λ. This sub-case fits the answer in Case 1 with the only difference that the stretching coefficients $\lambda_\pm$ are equal.

Otherwise — all eigenvectors of T are proportional to one of them which we denote $\mathbf{v}_+$. We have $T\mathbf{v}_+ = \lambda\mathbf{v}_+$ where $\mathbf{v}_+ \neq \mathbf{0}$. We claim that there exists a vector $\mathbf{v}_-$ non-proportional to $\mathbf{v}_+$ such that $T\mathbf{v}_- = \lambda\mathbf{v}_- + \mathbf{v}_+$.

Let us examine first the subcase $\lambda = 0$. The entire line containing $\mathbf{v}_+$ is then transformed by T to $\mathbf{0}$. Since $T \neq \mathbf{0}$, the range of T is some line on the plane. This line consists of eigenvectors of T, and therefore the must coincide with the line containing $\mathbf{v}_+$. Thus $\mathbf{v}_+$ is in the range of T, and we can take on the role of $\mathbf{v}_-$ any vector such that $T\mathbf{v}_- = \mathbf{v}_+$. In the basis $\mathbf{v}_+, \mathbf{v}_-$ the linear transformation acts as the coordinate shift $\mathbf{v}_- \mapsto \mathbf{v}_+ \mapsto \mathbf{0}$ and has the matrix $\begin{bmatrix} 0 & 1 \\ 0 & 0 \end{bmatrix}$.

Let us consider now the general case. The characteristic polynomial of the non-zero matrix $T - \lambda I$ has the double root 0. Constructing the basis $\mathbf{v}_+, \mathbf{v}_-$ as in the previous subcase we see that $T - \lambda I$ acts on $\mathbf{v}_+, \mathbf{v}_-$ as the coordinate shift, and therefore T transforms $\mathbf{v}_-$ to $\lambda\mathbf{v}_- + \mathbf{v}_+$ and $\mathbf{v}_+$ — to $\lambda\mathbf{v}_+$. Thus in this basis T has the matrix $\begin{bmatrix} \lambda & 1 \\ 0 & \lambda \end{bmatrix}$ which is called the Jordan cell.

Examples. (a) When $\lambda = 1$ the matrix transforms rectangles aligned with coordinate axes to parallelograms of the same altitude. This transformation T is called shear.

(b) The transformation T with $\lambda \neq 0$ in the modified basis $\mathbf{v}_1 = \mathbf{v}_+$, $\mathbf{v}_2 = \lambda\mathbf{v}_-$ has the matrix $\lambda\begin{bmatrix} 1 & 1 \\ 0 & 1 \end{bmatrix}$ which describes the composition of the shear with multiplication by λ.

Let us summarize our results in both matrix and geometrical and forms.

Theorem (Similarity classification of matrices).
Any real 2×2-matrix is similar to exactly one of the following normal forms

$$\begin{bmatrix} \lambda_+ & 0 \\ 0 & \lambda_- \end{bmatrix}, \begin{bmatrix} a & -b \\ b & a \end{bmatrix}, \begin{bmatrix} \lambda & 1 \\ 0 & \lambda \end{bmatrix},$$

with respectively real eigenvalues $\lambda_+ \geq \lambda_-$, complex conjugate eigenvalues $a \pm bi$, $b > 0$, and a single real eigenvalue λ.

Theorem $'$ (Classification of linear transformations).
Any linear transformation on the plane in a suitable coordinate system is a coordinate-wise stretch (possibly with different coefficients) or a complex multiplication by $a + bi = re^{i\theta}$, or the composition of multiplication by a real $\lambda \neq 0$ with the shear, or the coordinate shift.

Remark. Generally speaking the basis vectors $\mathbf{v}_\pm$ of the coordinate system in question have no reason to be orthogonal. In particular, the directions of the eigenvectors in Case 1 and the rotation angle θ in Case 2 may have little in common with those in our description of invertible linear transformations as compositions of stretching and rotation given in Example 1.3.3(b).

Examples. (c) The Hamilton – Cayley identity: *any square matrix A satisfies its characteristic equation, $A^2 - (\operatorname{tr} A)A + (\det A)I = \mathbf{0}$.*

Indeed, if $A = \Lambda = \begin{bmatrix} \lambda_1 & 0 \\ 0 & \lambda_2 \end{bmatrix}$ is diagonal, then the characteristic polynomia equals $(\lambda - \lambda_1)(\lambda - \lambda_2)$, and it is easy to check that $(\Lambda - \lambda_1 I)(\Lambda - \lambda_2 I) = \mathbf{0}$. This equality remains true even if Λ is the Jordan cell with the eigenvalue $\lambda_1 = \lambda_2$.

In general A is similar to a (complex) diagonal matrix or a Jordan cell: $A = C\Lambda C^{-1}$. For any polynomial $P(\lambda)$ we have $P(A) = CP(\Lambda)C^{-1}$ since $A^k = C\Lambda C^{-1}C\Lambda C^{-1}...\Lambda C^{-1} = C\Lambda^k C^{-1}$. In particular $(A - \lambda_1 I)(A - \lambda_2 I) = C(\Lambda - \lambda_1 I)(\Lambda - \lambda_2 I)C^{-1} = \mathbf{0}$. The Hamilton – Cayley identity follows now from the fact that similar matrices A and Λ have the same characteristic polynomial.

(d) The matrix $A = \begin{bmatrix} 1 & -3 \\ 1 & 1 \end{bmatrix}$ has the characteristic polynomial $\lambda^2 - 2\lambda + 4$. Thus $A^2 = 2A - 4I$, $A^3 = 2A^2 - 4A = -8I$, and $A^{1999} = (-8)^{666}A = 2^{1998}A$.

(e) *Linear recursive sequences and dynamical systems.*

The Fibonacci sequence $0, 1, 1, 2, 3, 5, 8, 13, ...$ is defined recursively by the rule that each next term equals the sum of the previous two terms. More generally, a sequence of numbers $s_0, s_1, ..., s_n, ...$ is called a linear recursive sequence (of order

2, with coefficients p, q) if it satisfies $s_{n+1} = ps_n + qs_{n-1}$ starting from $n = 1$. Given two consecutive terms (s_{n-1}, s_n) of the sequence, the recursion relation allows to find the next pair of consecutive terms $(s_n, s_{n+1}) = (s_n, qs_{n-1} + ps_n)$, then — by the same formulas — the next pair, etc. We may think of the recursion relation as a **dynamical system**: at each discrete count of time $n = 1, 2, 3, ...$ the current state of the system described by the vector $\mathbf{x}_{n-1} = \begin{bmatrix} s_{n-1} \\ s_n \end{bmatrix}$ jumps to $\mathbf{x}_n = \begin{bmatrix} s_n \\ s_{n+1} \end{bmatrix}$. All possible states of the system (that is all possible pairs of consecutive terms) form the plane called the **phase plane** of the dynamical system. The dynamical system is linear in the sense that the new position is obtained from the old one by the linear transformation $\mathbf{x}_n = A\mathbf{x}_{n-1}$ with the matrix $A = \begin{bmatrix} 0 & 1 \\ q & p \end{bmatrix}$. Trajectories $(\mathbf{x}_0, \mathbf{x}_1, ..., \mathbf{x}_n, ...)$ of the dynamical system on the phase plane are uniquely determined by the initial positions $\mathbf{x}_0$ since $\mathbf{x}_n = A^n\mathbf{x}_0$ and, if the matrix A is invertible, the "past history" is unambiguously recovered from the current state too: $\mathbf{x}_{-n} = A^{-n}\mathbf{x}_0$. Linear dynamical systems more general than those corresponding to recursive sequences can be constructed by picking an arbitrary invertible matrix on the role of A.

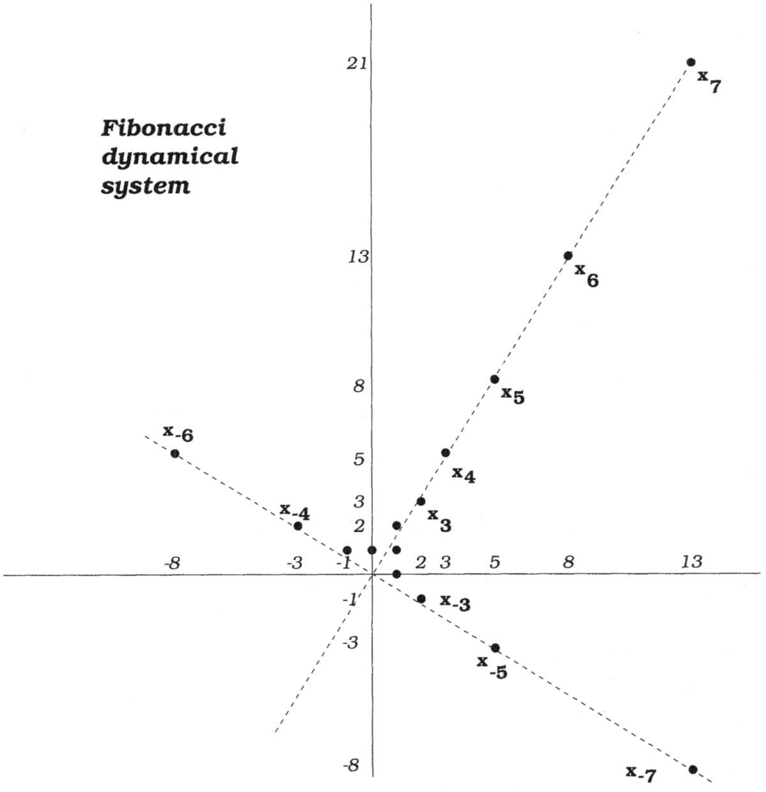

Normal forms of linear transformations allow one to analyze the behavior of trajectories in terms of eigenvectors and eigenvalues. The characteristic polynomial of the matrix A equals $\lambda^2 - p\lambda - q$. Suppose that the roots $\lambda_\pm$ of the characteristic polynomial are distinct, and $\mathbf{v}_\pm$ are corresponding eigenvectors (real or complex).

Then the trajectory with the initial condition $\mathbf{x}_0 = \alpha_+\mathbf{v}_+ + \alpha_-\mathbf{v}_-$ can be computed as $\mathbf{x}_n = A^n\mathbf{x}_0 = \alpha_+\lambda_+^n\mathbf{v}_+ + \alpha_-\lambda_-^n\mathbf{v}_-$. For instance, if the eigenvalues are real the formula describes the behavior of the discrete trajectories on the phase plane as superposition of stretching with the rate λ_+ in the direction of the first eigenvector and with the rate λ_- in the direction of the second eigenvector.

By picking the first row in the above vector equality we obtain a "general formula" for terms of the recursive sequence $s_n = a_+\lambda_+^n + a_-\lambda_-^n$ where the coefficients $a_\pm$ are to be found from the initial condition (s_0, s_1). For example, the Fibonacci sequence corresponds to the characteristic polynomial $\lambda^2 - \lambda - 1$ with the roots $\lambda_\pm = (1 \pm \sqrt{5})/2$. For $n = 0, 1$ we have $a_+ + a_- = s_0 = 0$, $a_+\lambda_+ + a_-\lambda_- = s_1 = 1$ and therefore $a_\pm = \pm 1/\sqrt{5}$. Thus the n-th Fibonacci number $s_n = \sqrt{5}^{-1} 2^{-n}[(1+\sqrt{5})^n - (1-\sqrt{5})^n]$.

Exercises 1.5.3.

(a) Find the place of the linear transformation $\begin{bmatrix} 2 & -4 \\ 1 & -2 \end{bmatrix}$ in the classification.

(b) Show that the following matrices are similar:

$$\begin{bmatrix} 1 & 1 \\ 4 & -2 \end{bmatrix}, \begin{bmatrix} 3 & 3 \\ -2 & -4 \end{bmatrix}.$$

(c) Compute $\begin{bmatrix} 3 & -7 \\ 1 & -2 \end{bmatrix}^{99}$.

(d) Diagonalize the following matrices by real or complex similarity transformations:

$$\begin{bmatrix} 1 & 0 \\ 1 & -1 \end{bmatrix}, \begin{bmatrix} 1 & -1 \\ 2 & -1 \end{bmatrix}.$$

(e) For the following recursive sequences, (i) sketch the first 7 points on the phase plane and (ii) find the general formula:

$$s_0 = 0, \; s_1 = 1, \; s_{n+1} = s_n - s_{n-1}, \; n > 1,$$

$$s_0 = 0, \; s_1 = 1, \; 2s_{n+1} = 3s_n + 2s_{n-1}, \; n > 1.$$

(f) Is there a 2×2-matrix A such that $A \neq \mathbf{0}$, but $A^2 = \mathbf{0}$? $A^2 \neq \mathbf{0}$ but $A^3 = \mathbf{0}$?

(g) Prove that any 2×2 matrices A, B, C satisfy the identity

$$(AB - BA)^2 C = C(AB - BA)^2.$$

(h) Consider linear transformations defined by traceless matrices $\begin{bmatrix} a & b \\ c & -a \end{bmatrix}$. In the a, b, c-space, draw the sets formed by linear transformations with only one real eigenvalue, no real eigenvalues, two real eigenvalues. Describe the partition of the a, b, c-space into similarity classes of traceless matrices.

SAMPLE MIDTERM EXAM

1. Formulate the definition of an orthogonal transformation and prove that a linear transformation preserving lengths of all vectors is orthogonal.

2. Find directions of principal axes of the quadratic form $Q(x_1, x_2) = 5x_1^2 + 12x_1x_2 + 10x_2^2$ and scketch the curve $Q(x_1, x_2) = 13$.

3. Is A similar to A^{1999}, if

$$A = \begin{bmatrix} 2 & -1 \\ 3 & -2 \end{bmatrix}, \; A = \begin{bmatrix} 2 & -1 \\ 3 & -1 \end{bmatrix}, \; A = \begin{bmatrix} 3 & -2 \\ 2 & -1 \end{bmatrix}?$$

CHAPTER 2

Differential Equations

2.1. ODE

Ordinary Differential Equations are mathematical models of various processes in astronomy, physics, chemistry, biology, ecology, etc. Similarly to the last example of the previous chapter, ODEs can be considered as dynamical systems, but with continuous time variable in contrast with the example where the time was discrete.

2.1.1. Existence and uniqueness of solutions. The equation of the form $\dot{x} = F(t, x)$ is called the first order ordinary differential equation. A solution to this equation is a function $x(t)$ on some interval of the time variable t which is differentiable and satisfies $dx(t)/dt = F(t, x(t))$. In geometrical terms, the graph of the solution $x(t)$ on the plane with coordinates (t, x) should everywhere have the slope prescribed by the given function $F(t, x)$.

Examples. (a) The Fundamental Theorem of Calculus says that solutions to the differential equation $\dot{x} = f(t)$ where f is a given continuous function are obtained by integration:

$$x(t) = C + \int_{t_0}^{t} f(\tau) d\tau.$$

Substituting $t = t_0$ we find that $C = x(t_0)$ has the meaning of the initial value of the solution at the moment t_0. The graphs of the solutions are obtained from each other by vertical shifts and thus fill in the (t, x)-plane without intersections.

(b) *Populational dynamics.* We can interpret the differential equation $\dot{x} = \lambda x$ as the following reproduction law for the amount $x(t)$ of, say, some bacteria: the rate of grows $\dot{x}(t)$ is proportional to the current amount of bacteria with the coefficient λ, the reproduction rate. The solutions to this equation have the form $x(t) = Ce^{\lambda t}$, where $C = x(0)$ is the initial population.

(c) Separable equations have the form $dx/dt = f(t)/g(x)$ and are called so because they are solved by the following separation of variables:

$$\int_{x_0}^{x} g(\xi) d\xi = \int_{t_0}^{t} f(\tau) d\tau.$$

For instance, the linear 1-st order homogeneous equation $\dot{x} = \lambda(t)x$ with the varying "reproduction rate" is separable:

$$\int_{t_0}^{t} \lambda(\tau) d\tau = \int_{x_0}^{x} \frac{d\xi}{\xi} = \ln|x(t)| - \ln|x_0|,$$

hence the solution with the initial value $x(t_0) = x_0$ is

$$x(t) = x_0 e^{\int_{t_0}^{t} \lambda(\tau) d\tau}.$$

(d) *Populational explosion.* The equation $\dot{x} = x^2$ describes reproduction in the population with the growth rate proportional to the number of pairs. Separating the variables, we find

$$\int^{t} d\tau = \int^{x} \frac{d\xi}{\xi^2} = -\frac{1}{x} + const,$$

or $x(t) = 1/(const - t)$. We conclude that any initial population $x(0) = 1/const > 0$ with this reproduction law explodes to infinity after a finite time interval.

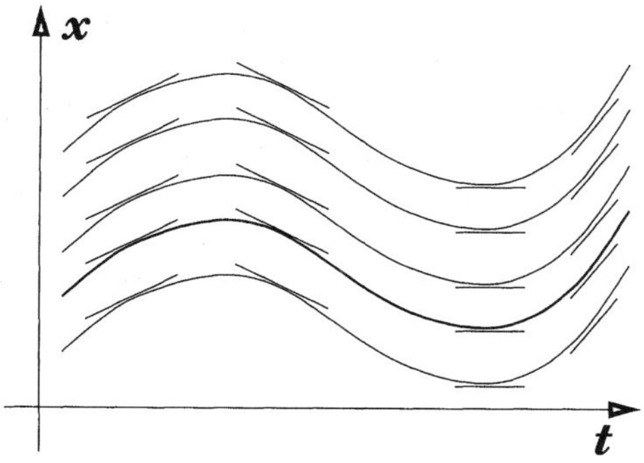

$\dot{x} = f(t)$

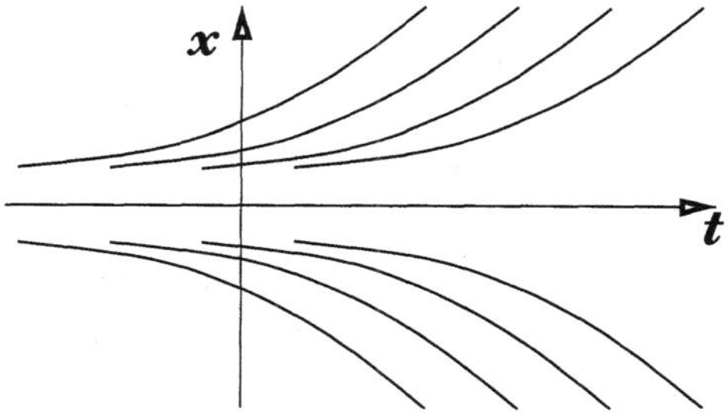

$\dot{x} = \lambda x$

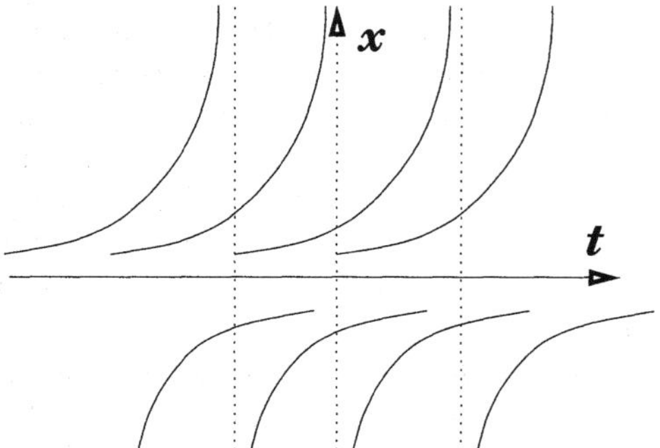

$\dot{x} = x^2$

The previous example shows that even if the right hand side of the differential equation is well-defined for all t, solutions do not have to exist on the entire time interval. However, in all the above examples a solution with the initial value $x(t_0) = x_0$ exists in some neighborhood of t_0 and is unambiguously determined in this neighborhood by this initial value. Geometrically this means that the domain of $F(t, x)$ is entirely filled in by the graphs of solutions without intersections. In fact the examples illustrate a general theorem that guarantees existence and uniqueness of solutions for systems of 1-st order ODEs

$$\begin{aligned} \dot{x}_1 &= F_1(t, x_1, ..., x_n) \\ &\dots \\ \dot{x}_n &= F_n(t, x_1, ..., x_n) \end{aligned}$$

under some mild hypotheses about the functions $F_1, ..., F_n$.

Theorem. *Suppose that the functions $F_1, ..., F_n$ have continuous derivatives in each of the variables $t, x_1, ..., x_n$. Then a solution $\mathbf{x}(t) = (x_1(t), ..., x_n(t))$ to the ODE system with the initial value $\mathbf{x}(t_0) = (x_1^0, ..., x_n^0)$ exists on some time interval containing t_0, and any two solutions with the same initial values coincide (on the common interval of their existence).*

Remark. A precise meaning of the differentiability hypothesis for functions in several variables will become more clear when we introduce partial derivatives. The hypothesis is actually stronger than necessary and can be relaxed. However it is not redundant, as the following counter-example shows.

Examples. (e) The differential equation $\dot{x}^3 = x^2$ has a solution $x(t) = t^3/27$ satisfying the same initial condition $x(0) = 0$ as the identically zero solution $x(t) = 0$. In fact the equation has infinitely many solutions (find them!) with the same initial value $x(0) = 0$. Of course, this does not contradict the theorem since the equation does not have the form $\dot{x} = F(t, x)$. However we can resolve it with respect to $\dot{x}$ and get the new equation $\dot{x} = (x^2)^{1/3}$ which has the required form and has infinitely many solutions with the same initial value. Yet this does not contradict the theorem since the function $x^{2/3}$ is not differentiable at $x = 0$.

(f) A 2-nd order ODE $\ddot{x} = G(t, x, \dot{x})$ can be transformed to a system of two 1-st order ODEs by the following standard trick. Put $x_1 = x$, $x_2 = \dot{x}$. Then the system $\dot{x}_1 = x_2$, $\dot{x}_2 = G(t, x_1, x_2)$ is equivalent to the original equation. Indeed, given a solution $x(t)$, the functions $x_1 = x(t), x_2 = dx(t)/dt$ satisfy the system, and vice versa, if $(x_1(t), x_2(t))$ is a solution to the system, then $x = x_1(t)$ satisfies the 2-nd order equation. The existence and uniqueness theorem applies to the system and says that if the function $G(t, x_1, x_2)$ has continuous derivatives then a solution is uniquely determined by the initial conditions $(t_0, x_1(t_0), x_2(t_0)) = (t_0, x(t_0), \dot{x}(t_0))$. In particular, we see that in order to specify a solution of a 2-nd order equation it is not sufficient to know the value $x(t_0)$ of the solution at the moment t_0 but it is also necessary to specify the velocity $\dot{x}(t_0)$ at this moment.

(g) Consider a pendulum of mass m suspended on a weightless rod of length l and swinging without friction under the influence of the gravity force. Denote by $x(t)$ the angle the rod makes with the downward direction of the vertical axis at the moment t. The component of the gravity force in the radial direction is compensated by the tension of the rod, while the component of the gravity vector in the tangential direction causes the pendulum to move along the circle of radius

l. The motion obeys the Newton equation

$$\text{mass} \times \text{acceleration} = \text{force}.$$

Accelerations are the 2-nd time derivatives of positions, and therefore the Newton equation gives rise to 2-nd order ODEs whenever the force can be described as a function of positions and/or velocities. In our situation it yields the equation of the mathematical pendulum

$$ml\ddot{x} = -mg\sin x\ , \text{ or } \ddot{x} = -\frac{g}{l}\sin x,$$

where $g \approx 9.8\ \mathtt{m/sec}^2$ is the gravitational acceleration. According to the existence and uniqueness theorem the angular position $x(0)$ and angular velocity $\dot{x}(0)$ at the moment $t = 0$ uniquely determine the future and the past motion $x(t)$ of the pendulum.

Assuming that the pendulum experiences air resistance proportional to the angular velocity we arrive to the equation of the damped pendulum

$$\ddot{x} = -k\dot{x} - \frac{g}{l}\sin x.$$

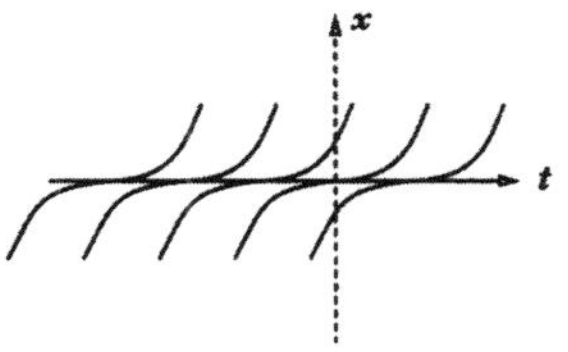

$\dot{x}^3 = x^2$

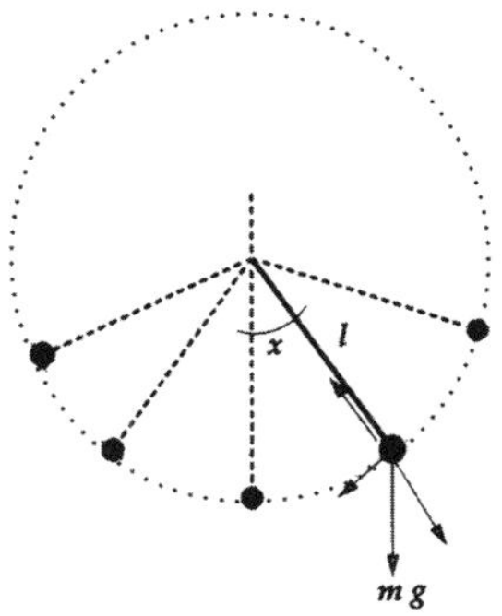

(h) *Phase portraits.* The equations of damped and undamped penduli have the form $\ddot{x} = G(x, \dot{x})$ with the right hand side independent on the time variable. Such equations are called time independent or autonomous. If $x(t)$ is a solution of an autonomous equation (satisfying some initial condition at $t = 0$) then $x(t-t_0)$ is also a solution of the same equation (satisfying the same initial condition at $t = t_0$). Such invariance to the time shift makes it convenient to represent solutions of autonomous equations graphically on the phase plane with coordinates $(x_1, x_2) = (x, \dot{x})$. The same applies to an autonomous system $\dot{x}_1 = F_1(x_1, x_2)$, $\dot{x}_2 = F_2(x_1, x_2)$. Namely, at each point of the plane with the radius-vector $\mathbf{x} = (x_1, x_2)$ the right hand side of the system specifies a vector $\mathbf{F}(\mathbf{x})$ with coordinates F_1, F_2. A solution

$\mathbf{x}(t) = (x_1(t), x_2(t))$ describes a point moving in the plane in such a way that the velocity vector $\dot{\mathbf{x}}(t)$ at each moment coincides with the prescribed vector $\mathbf{F}(\mathbf{x})$: $\dot{\mathbf{x}}(t) = \mathbf{F}(\mathbf{x}(t))$. The solutions $\mathbf{x}(t)$ and $\mathbf{x}(t - t_0)$, obtained from one another by a time shift, describe the same trajectories on the plane. The uniqueness and existence theorem guarantees that the trajectories described by solutions fill in the plane entirely and that any two trajectories having a common point actually coincide in which case the corresponding solutions differ only by a time shift. The trajectories are called phase curves.

(i) A time dependent equation $\dot{x} = F(t, x)$ can be always rewritten as a time-independent system using the following simple (and almost useless) trick: put $x_1 = t$, $x_2 = x$; then the system $\dot{x_1} = 1$, $\dot{x_2} = F(x_1, x_2)$ is equivalent to the original equation. In particular, the graphs of solutions in Examples (a,b,d,e) can be considered as phase curves of the corresponding system.

Remark. We are not going to prove the existence and uniqueness theorem and will seldom refer to it. It is useful however to have the theorem in mind and to understand its fundamental role. The existence statement of the theorem is usually intuitively obvious in applications of ODEs to ecology, astronomy, physics, etc. since it simply says that the process described by the differential equation *goes* (populations change, planets revolve, penduli swing, water flows). Thus the mathematical theorem here confirms our intuition and shows that the models based on ODEs have a chance to describe real phenomena adequately. The uniqueness statement of the theorem shows however that ODEs can describe only deterministic phenomena, that is only those processes where the future and the past is unambiguously determined by the current state of the system. For each particular system it also tells us how many quantities is needed in order to specify a current state.

2.1.2. Linear ODE systems. It happens suspiciously often that empirical "laws of nature" assume the form of linear proportionality. The Ohm Law says that the electric current through a resistor is proportional to the voltage. According to the Hooke Law, the length increment Δl of an elastic rod or spring caused by a stretching force is proportional to the force. The thermal expansion coefficient of a given material describes dimensional variations caused by (and proportional to) variations of temperature.

On the other hand, you can find in physics handbooks graphical representations of thermal expansion coefficients as functions of the temperature (meaning that the thermal expansion is not really proportional to the temperature variations). A realistic dependence of the stretching force σ as a function of elongation Δl up to the breaking point of the rod is shown on the picture and is highly non-linear too. The most of real phenomena is actually non-linear. However many of them can be described by differentiable functions. The amazing persistence of linear proportionality laws in fundamental sciences is probably due to the fact that differentiable functions can be approximated by linear ones on sufficiently small intervals. Linear differential equations usually arise as a result of such approximation.

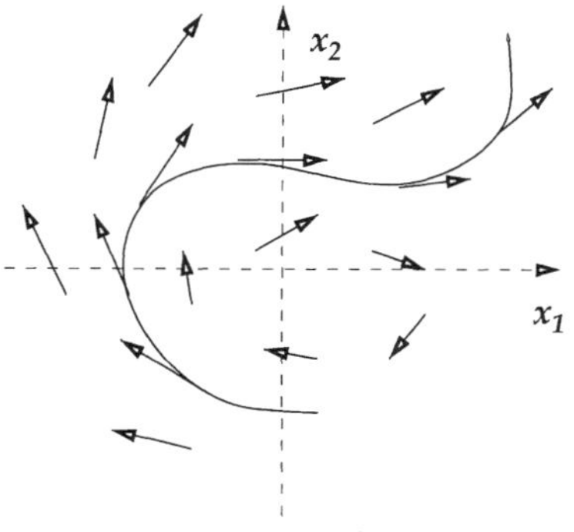

A phase curve of $\dot{x} = F(x)$

" *The Hooke Law* "

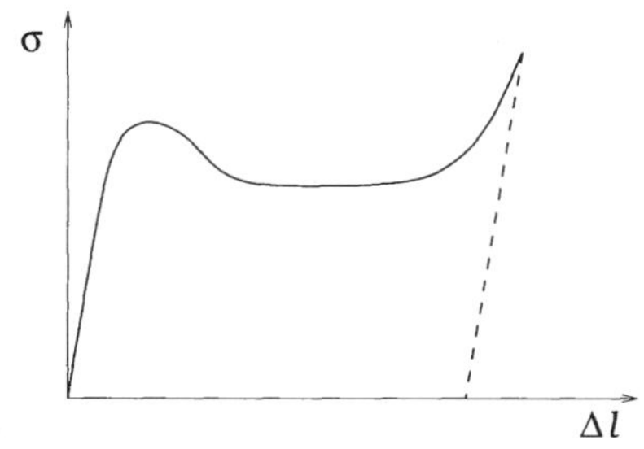

Exercises 2.1.1.

(a) Solve the following differential equations, sketch graphs of a few solutions for each of them and find the solutions satisfying given initial values:

$$\dot{x} + tx = 0, \ x(0) = 1;$$

$$t\dot{x} + 4x = 0, \ x(1) = 1;$$

$$(t^2 - 1)\dot{x} + 2tx^2 = 0, \ x(0) = 1;$$

$$\dot{x} \cot t + x = 2, \ x(0) = -1;$$

$$\dot{x} = 3(x^2)^{1/3}, \ x(2) = 0;$$

$$t\dot{x} + x = x^2, \ x(1) = 0.5.$$

(b) Consider the differential equation $\dot{x} = x(k - x)$, $k > 0$, as a model of reproduction of salmon with the reproduction rate $\lambda = k - x$ decreasing as a function of the population x due to limited food supply. Describe evolution of the population x in time if the initial value $0 < x(0) < k$, $x(0) = k$, $x(0) > k$.

(c) The population of salmon from Problem (b) is now subject to harvesting with a constant quota c. Study the populational dynamics described by the differential equation $\dot{x} = x(k - x) - c$. Find the limit of the population $x(t)$ when $t \to \infty$. For which values c of the harvest quota the population will extinct? survive? Find the quota c which maximizes the harvest in a long-term fishery?

(d) For the pendulum equation $\ddot{x} + \sin x = 0$, prove the energy conservation law: *if $x(t)$ is a solution, then*

$$E = \frac{\dot{x}^2}{2} - \cos x(t)$$

does not depend on t. For solutions with the energy $E = -1, -1/2, 0, 1/2, 1, 2$, find the maximal magnitude x of oscillations and the velocity $\dot{x}$ at the moment when the pendulum passes the position $x = 0$. Sketch the phase curves on the phase plane with coordinates $x, \dot{x}$ corresponding to solutions with these energies. Which of the solutions are periodic?

Example. (a) Using the approximation $\sin x \approx x$ valid for small $|x|$ we arrive to the equation of the harmonic oscillator describing small oscillations of the mathematical pendulum near the equilibrium $x = 0$:

$$\ddot{x} = -\omega^2 x, \quad \text{where } \omega^2 = g/l.$$

In fact all elementary oscillatory systems, such as mass – spring systems or LC-circuits, are described by this ODE. It can be rewritten as a system of two 1-st order ODEs $\dot{x}_1 = \omega x_2, \ \dot{x}_2 = -\omega x_1$.

In general, a system of two 1-st order linear ODEs

$$\begin{array}{rcrcr} \dot{x}_1 & = & a_{11}x_1 & + & a_{12}x_2 \\ \dot{x}_2 & = & a_{21}x_1 & + & a_{22}x_2 \end{array}$$

is determined by the matrix A of its coefficients and can be rewritten in matrix notations as $\dot{\mathbf{x}} = A\mathbf{x}$.

Linear combinations of solutions to a linear system are solutions too:

If $\mathbf{x}^{(1)}(t)$ and $\mathbf{x}^{(2)}(t)$ are two solutions to the system then $\mathbf{x}(t) = c_1\mathbf{x}^{(1)}(t) + c_2\mathbf{x}^{(2)}(t)$ satisfies $\dot{\mathbf{x}} = c_1\dot{\mathbf{x}}^{(1)} + c_2\dot{\mathbf{x}}^{(2)} = c_1 A\mathbf{x}^{(1)} + c_2 A\mathbf{x}^{(2)} = A(c_1\mathbf{x}^{(1)} + c_2\mathbf{x}^{(2)}) = A\mathbf{x}$. This conclusion remains true even if the matrix entries a_{ij} (but not c_1, c_2!) depend on t.

Example. (b) The functions $\cos\omega t$ and $\sin\omega t$ satisfy the harmonic oscillator equation $\ddot{x} = -\omega^2 x$. Thus $x(t) = c_1\cos\omega t + c_2\sin\omega t$ also satisfy it. On the phase plane with the coordinates $x_1 = x, \ x_2 = \dot{x}/\omega$ the arbitrary coefficients $c_1 = x_1(0), c_2 = x_2(0)$ represent the initial value of the solution at $t = 0$. According to the existence and uniqueness theorem all solutions to the harmonic oscillator equation are therefore the linear combinations of $\cos\omega t$ and $\sin\omega t$.

We will assume further on that the linear ODE system $\dot{\mathbf{x}} = A\mathbf{x}$ is time independent so that the matrix A has constant coefficients. A linear change of coordinates $\mathbf{x} = C\mathbf{y}$ on the phase plane transforms the system to $\dot{\mathbf{y}} = C^{-1}\dot{\mathbf{x}} = C^{-1}A\mathbf{x} = C^{-1}AC\mathbf{y}$ with the coefficient matrix $C^{-1}AC$ similar to A. We will apply our classification of matrices up to similarity transformations in order to simplify constant coefficient linear systems, solve them and describe their phase portraits.

Case 1: *Distinct real eigenvalues* λ_1, λ_2. The matrix A is similar to the diagonal matrix. The ODE system in the diagonalizing coordinates has the form $\dot{y}_1 = \lambda_1 y_1, \ \dot{y}_2 = \lambda_2 y_2$ of two independent 1-st order equations which is easy to solve:

$$y_1(t) = e^{\lambda_1 t}y_1(0), \quad y_2(t) = e^{\lambda_2 t}y_2(0).$$

Since $\mathbf{y}(0) = C^{-1}\mathbf{x}(0)$, we have

$$\mathbf{x}(t) = C\mathbf{y}(t) = C\begin{bmatrix} e^{\lambda_1 t} & 0 \\ 0 & e^{\lambda_2 t} \end{bmatrix} C^{-1}\mathbf{x}(0),$$

where C is the matrix of eigenvectors of A. In order to sketch the phase curves in the coordinate system (y_1, y_2) we eliminate t from the solution formulas: $\lambda_1 t = \ln y_1 + const$ and hence $y_2 = Const \cdot y_1^{\lambda_2/\lambda_1}$. Thus the phase curves are graphs of the power functions. The possible phase portraits with $\lambda_1, \lambda_2 \neq 0$ are shown on the picture and have the names: nodal sink for $0 > \lambda_1 > \lambda_2$, saddle for $\lambda_1 > 0 > \lambda_2$ and nodal source for $\lambda_1 > \lambda_2 > 0$. [1]

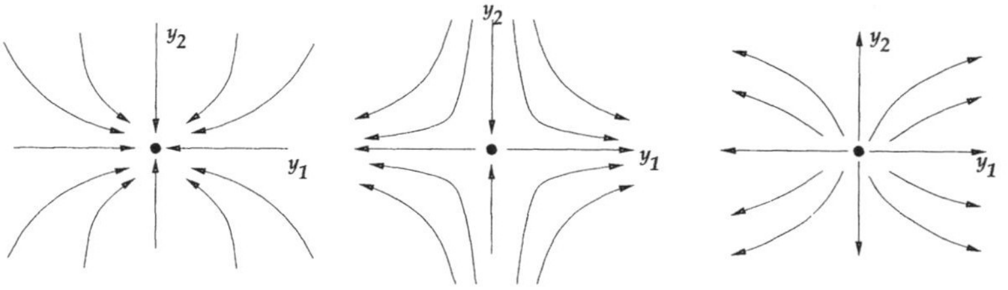

Nodal sink *Saddle* *Nodal source*

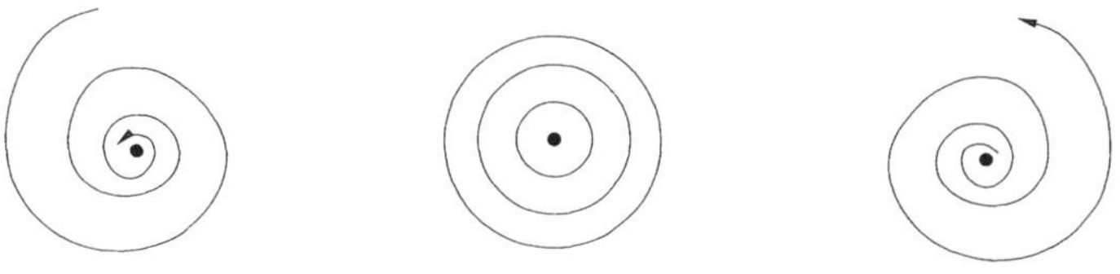

Spiral sink *Center* *Spiral source*

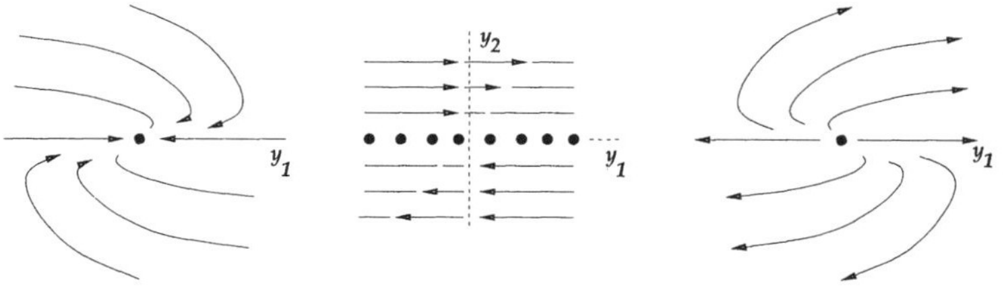

Degenerate sink *Free particle* *Degenerate source*

Case 2: Complex conjugate eigenvalues. Diagonalizing the matrix A by a complex similarity transformation we arrive to the same formulas as in the first case where however $\lambda_1 = \rho + i\omega, \lambda_2 = \rho - i\omega$ and $c_1 = a + ib,\ c_2 = a - ib$ are complex numbers, and columns of the matrix C represent complex conjugate eigenvectors. Purely real formulas for solutions can be extracted from these complex formulas by taking the real part. For instance, in the real coordinates $(u_1, u_2) = (\operatorname{Re} y_1, \operatorname{Im} y_1)$ on the phase plane the solutions to the ODE system read:

$$u_1 = \operatorname{Re}[e^{(\rho+i\omega)t}(a + ib)] = e^{\rho t}(a \cos \omega t - b \sin \omega t)$$

$$u_2 = \operatorname{Im}[e^{(\rho+i\omega)t}(a + ib)] = e^{\rho t}(a \sin \omega t + b \cos \omega t).$$

[1] The phase pictures on the (x_1, x_2)-plane are obtained from these by invertible linear transformations. As we know, such a transformation is a composition of a rotation with stretching in two perpendicular directions (which may have nothing to do with the directions of the eigenvectors). Thus typical phase portraits of linear systems with distinct non-zero real eigenvalues are "disturbed" nodal sources, saddles and nodal sinks.

Thus in the vector form $\mathbf{u}(t) = e^{\rho t}T_{\omega t}\mathbf{u}(0)$ where $T_{\omega t}$ are rotation matrices and the initial condition $\mathbf{u}(0) = (a, b)$. We conclude that the motion on the phase plane is described as the simultaneous rotation with the angular velocity ω and dilation with the rate ρ. The phase portraits corresponding to $\rho < 0$, $\rho = 0$ and $\rho > 0$ are called respectively spiral sink, center and spiral source.

Case 3: Multiple eigenvalue. When the matrix A is scalar, the ODE system has the form $\dot{x}_1 = \lambda x_1$, $\dot{x}_2 = \lambda x_2$, the solutions $x_1(t) = x_1(0)e^{\lambda t}$, $x_2(t) = x_2(0)e^{\lambda t}$ and the phase curves $x_2 = Const \cdot x_1$. The phase portraits with $\lambda > 0$ and $\lambda < 0$ are special examples of the nodal source and sink described in Case 1. Excluding the possibility that A is scalar, we transform the ODE system to the form $\dot{y}_1 = \lambda y_1 + y_2$, $\dot{y}_2 = \lambda y_2$ corresponding to $C^{-1}AC$ being the Jordan cell. If $\lambda = 0$ the system is equivalent to the Newton equation $\ddot{y} = 0$ of a free particle (which according to the Galilean Inertia Law moves with a constant velocity) and is easy to solve: $y_2(t) = y_2(0)$, $y_1(t) = y_2(0)t + y_1(0)$. The phase portrait of the free particle is shown on the picture. When $\lambda \neq 0$, the solutions are modified by the dilation with the rate λ:

$$y_1(t) = e^{\lambda t}(y_1(0) + ty_2(0)), \quad y_2(t) = e^{\lambda t}y_2(0).$$

The phase portraits with $\lambda > 0$ (degenerate source) and $\lambda < 0$ (degenerate sink) are modified accordingly. Solutions of the original ODE system are given by the formula

$$\mathbf{x}(t) = e^{\lambda t}C \begin{bmatrix} 1 & t \\ 0 & 1 \end{bmatrix} C^{-1}\mathbf{x}(0) \, .$$

Examples. We will study the behavior of the damped pendulum described near the lower equilibrium by the linear equation $\ddot{x} = -k\dot{x} - gx/l$. The coefficient matrix of the corresponding linear system

$$\begin{array}{rcrr} \dot{x}_1 & = & & x_2 \\ \dot{x}_2 & = & -\frac{g}{l}x_1 & -kx_2 \end{array}$$

has the characteristic polynomial $\lambda^2 + k\lambda + g/l$ with the roots

$$\lambda_\pm = -\frac{k}{2} \pm \sqrt{\frac{k^2}{4} - \frac{g}{l}}.$$

Properties of solutions change when the damping coefficient k grows, and we will analyze several cases and — at the same time — illustrate several practical techniques for solving linear ODE systems. [2]

(c) If $k = 0$, the eigenvalues $\lambda_\pm = \pm i\omega$, $\omega^2 = g/l$, are purely imaginary. It is the case of the harmonic oscillator. The phase portrait is a center, and the solutions

---

[2]These techniques can be briefly described as follows: in the example (e) we write down two particular solutions corresponding to two real eigenvectors and describe the general solution as their linear combination with arbitrary real coefficients; in the example (d) we take arbitrary linear combinations of the real and imaginary parts of a single complex solution corresponding to a complex eigenvector; in the example (f) we look for solutions in the form of linear combinations of suitable functions and extract the actual solutions by plugging the functions into the differential equations. We advise the reader to try this approach in the previous examples too.

$x_1 = x_1(0)\cos\omega t + \omega^{-1}x_2(0)\sin\omega t,\ x_2 = -\omega x_1(0)\sin\omega t + \omega x_2(0)\cos\omega t$ are $2\pi/\omega$-periodic. Notice that the period of oscillations does not depend on the magnitude of oscillations of the pendulum. [3]

(d) If $0 < k < \sqrt{4g/l}$, the eigenvalues $\lambda_\pm = \rho \pm i\omega$ are complex conjugate and have the negative real part $\rho = -k/2$. The phase portrait is a spiral sink. In order to find the solutions, we first find a complex eigenvector $\mathbf{v}$ with the eigenvalue, say, λ_+: $\mathbf{v} = (1, \lambda_+)$, then notice that $\mathbf{z}(t) = e^{\lambda_+ t}\mathbf{v}$ is a complex solution to the system and conclude that the real and imaginary parts of this solution are two real solutions to the system:

$$\begin{array}{rclrcl} x_1^{(1)} & = & e^{\rho t}\cos\omega t & x_1^{(2)} & = & e^{\rho t}\sin\omega t \\ x_2^{(1)} & = & e^{\rho t}(\rho\cos\omega t - \omega\sin\omega t) & x_2^{(2)} & = & e^{\rho t}(\rho\sin\omega t + \omega\cos\omega t) \end{array}.$$

The general solution is a linear combination of these two particular solutions with arbitrary coefficients: $\mathbf{x}(t) = c_1\mathbf{x}^{(1)}(t) + c_2\mathbf{x}^{(2)}$. In particular oscillations of the angular coordinate $x = x_1$ of the damped pendulum are described by $x(t) = e^{\rho t}(c_1\cos\omega t + c_2\sin\omega t) = x(t_0)e^{-kt/2}\cos(\omega(t - t_0))$. The cosine factor exhibits the oscillatory character of the motion, but the frequency $\omega = \sqrt{g/l - k^2/4}$ is smaller than in the case of undamped pendulum. The exponential factor indicates that the magnitude of oscillations decreases with time.

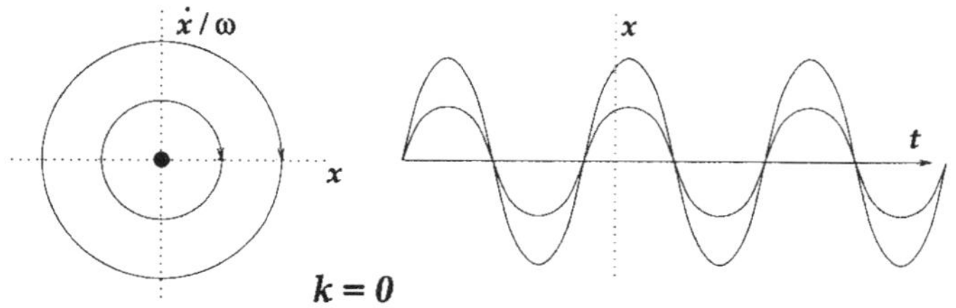

$k = 0$

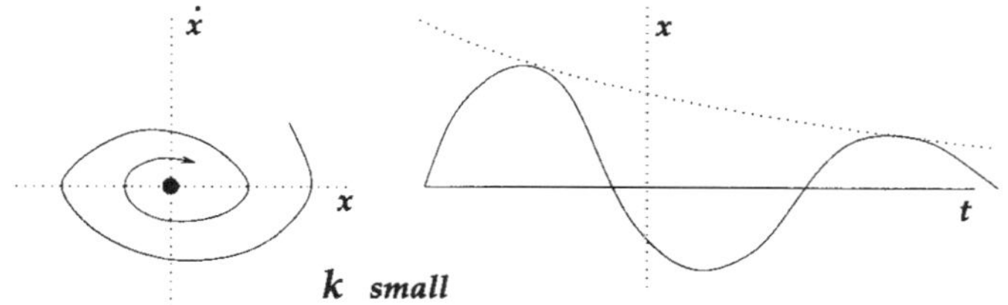

k small

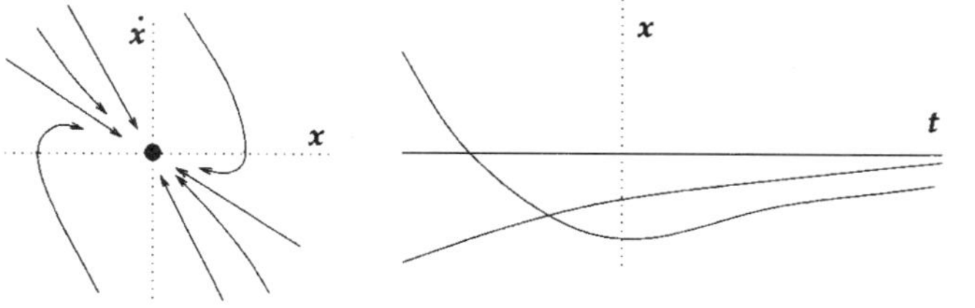

k large

[3]This observation used to serve as a mathematical foundation for designing the pendulum clock invented by Galileo Galilei (1564 – 1642).

(e) If $k > \sqrt{4g/l}$, the eigenvalues $\lambda_\pm$ are real, negative and distinct. The real eigenvectors $\mathbf{v}_\pm = (1, \pm\lambda)$ correspond to two particular solutions $\mathbf{x}_\pm(t) = e^{\lambda_\pm t}\mathbf{v}_\pm$, and the general solution is their linear combination

$$\begin{array}{rcll} x_1(t) & = & c_+ e^{\lambda_+ t} & + \quad c_- e^{\lambda_- t} \\ x_2(t) & = & c_+\lambda_+ e^{\lambda_+ t} & + \quad c_-\lambda_- e^{\lambda_- t} \end{array}.$$

The phase portrait is a nodal sink, and the direction of the eigenlines on the phase plane are specified by the vectors $\mathbf{v}_\pm$. We conclude that in the case of strong damping all solutions are aperiodic and decay exponentially with time. Moreover, the pendulum passes the vertical position at most once (since each phase curve on the portrait crosses the line $x_1 = 0$ no more than once).

(f) In the borderline case $k = \sqrt{4g/l}$ the eigenvalue $\lambda = -k/2$ is multiple and negative. The matrix of the system is similar to the Jordan cell (since the matrix is not scalar). The two eigenlines of the previous example merge to a single one with the slope $-k/2$. This explains the phase portrait shown on the picture. In order to find explicit solution formulas, let us notice that in the Jordan cell case both components of solutions are linear combinations of functions $e^{\lambda t}$ and $te^{\lambda t}$. Thus the solution has the form $x_1 = e^{\lambda t}(c_1 + c_2 t)$, $x_2 = e^{\lambda t}(c_3 + c_4 t)$. This leaves 4 arbitrary constants (instead of 2). Substituting the formulas into the differential equations we will therefore find two relations among $c_1, \ldots, c_4$. The equation $\dot{x}_1 = x_2$ yields $\lambda c_1 + c_2 = c_3$ and $\lambda c_2 = c_4$. Thus the general solution to the system is

$$x_1(t) = e^{-kt/2}(c_1 + c_2 t), \; x_2(t) = e^{-kt/2}(c_2 - \frac{k}{2}c_1 - \frac{k}{2}c_2 t).$$

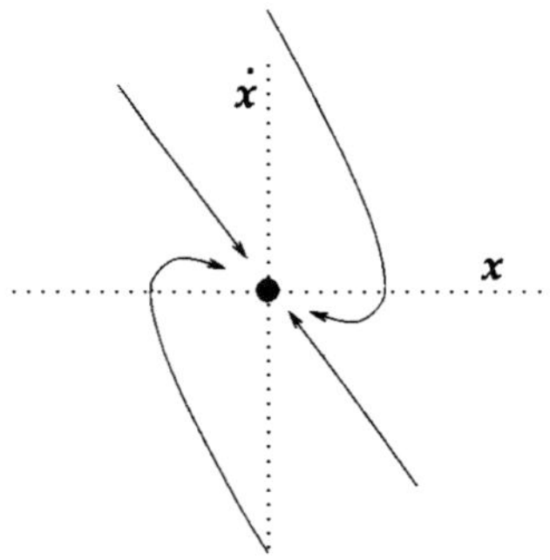

$k^2 = 4gl$

Remark. The damped harmonic oscillator equation we have just studied is an example of 2-nd order constant coefficient ODEs $\ddot{x} + p\dot{x} + qx = 0$. In order to find solutions to such an equation it suffices to notice that the matrix $\begin{bmatrix} 0 & 1 \\ -q & -p \end{bmatrix}$ of the corresponding system has the characteristic polynomial $\lambda^2 + p\lambda + q$. Solutions to the equation will therefore have the form $x(t) = c_1 e^{\lambda_1 t} + c_2 e^{\lambda_2 t}$ if the roots of the polynomial are real and distinct, $x(t) = e^{\rho t}(c_1 \cos\omega t + c_2 \sin\omega t)$ if the roots $\lambda_\pm = \rho \pm \omega$ are complex conjugate, and $x(t) = e^{\lambda t}(c_1 + c_2 t)$ in the case of multiple roots.

Exercises 2.1.2.

(a) Solve the following linear ODE systems, sketch the phase portraits on the phase plane with coordinates (x_1, x_2), find the solutions with the initial values $x_1(0) = 1$, $x_2(0) = 0$ and sketch the graphs of $x_1(t)$ and $x_2(t)$:

$$\begin{matrix} \dot{x}_1 = 3x_1 \\ \dot{x}_2 = 2x_1 + x_2 \end{matrix}, \quad \begin{matrix} \dot{x}_1 = x_1 + 2x_2 \\ \dot{x}_2 = 5x_2 - 2x_1 \end{matrix}, \quad \begin{matrix} \dot{x}_1 = x_1 + 3x_2 \\ \dot{x}_2 = -6x_1 - 5x_2 \end{matrix}, \quad \begin{matrix} \dot{x}_1 = x_1 \\ \dot{x}_2 = 2x_1 - x_2 \end{matrix},$$

$$\begin{matrix} \dot{x}_1 = -2x_1 - 5x_2 \\ \dot{x}_2 = 2x_1 + 2x_2 \end{matrix}, \quad \begin{matrix} \dot{x}_1 = 3x_1 + x_2 \\ \dot{x}_2 = x_2 - x_1 \end{matrix}, \quad \begin{matrix} \dot{x}_1 = 3x_1 - 2x_2 \\ \dot{x}_2 = 4x_2 - 6x_1 \end{matrix}, \quad \begin{matrix} \dot{x}_1 = x_2 - 2x_1 \\ \dot{x}_2 = 2x_2 - 4x_1 \end{matrix}.$$

(b) Sketch the graphs of solutions $x(t)$ to the damped pendulum equation with $k = 2\sqrt{g/l}$ corresponding to the 5 phase curves shown on the phase portrait.

(c) Find the linear 2-nd order ODE describing the behavior of the undamped pendulum near the upper equilibrium ($x \approx \pi$) and sketch the phase portrait of the equation on the phase plane with coordinates $(x, \dot{x})$. Find those solutions $x(t)$ to the equation which tend to the equilibrium when $t \to \infty$. Which motions of the non-linear pendulum are approximated by these solutions?

(d) Consider periodic solutions of the non-linear undamped pendulum equation $\ddot{x} + \sin x = 0$. Does the period of oscillations depend on the magnitude?

2.2. Stability

Linear ODE systems usually arise as approximations to non-linear systems near equilibria. The linear approximations are often sufficient in order to judge whether a small deviation from the equilibrium will cause solutions to drift far away from it or the equilibrium will be reestablished automatically. In practical applications this issue translates into the crucial question: will, say, a chemical reactor blow up or work steadily near the intended regime? Before explaining how to analyze stability of equilibria we have to discuss linear approximations of functions.

2.2.1. Partial derivatives. Let $y = f(\mathbf{x})$ be a function of two variables $\mathbf{x} = (x_1, x_2)$.

The function is called continuous at a point $\mathbf{a} = (a_1, a_2)$ if it can be approximated by a constant function with an error smaller than any constant function in a sufficiently small neighborhood of $\mathbf{a}$. The phrase "smaller than any constant function" sounds ambiguous and needs explanations. In order to make the error at $\mathbf{x} = \mathbf{a}$ smaller than any constant we must take the value $f(\mathbf{a})$ on the role of the approximating constant. Then the error function $f(\mathbf{x}) - f(\mathbf{a})$ vanishes at $\mathbf{x} = \mathbf{a}$. We say that "the error is smaller than any constant" if for any positive constant ε the error function becomes smaller than ε in a sufficiently small disk around $\mathbf{a}$. In other words, f is continuous at $\mathbf{a}$ if $\lim_{\mathbf{x}\to\mathbf{a}} |error(\mathbf{x})| = 0$.

The function f is called differentiable at $\mathbf{a}$ if it can be approximated by a linear function with an error smaller than any linear function in a sufficiently small neighborhood of $\mathbf{a}$. Since linear functions grow in some directions proportionally to the distance $|\mathbf{x} - \mathbf{a}|$, the previous sentence should be interpreted in the following way: for some α, β the error $f(\mathbf{x}) - f(\mathbf{a}) - [\alpha(x_1 - a_1) + \beta(x_2 - a_2)]$ of the linear approximation satisfies $\lim_{\mathbf{x}\to\mathbf{a}} \frac{|error(\mathbf{x})|}{|\mathbf{x}-\mathbf{a}|} = 0$.

The function f is called two times differentiable at $\mathbf{a}$ if it can be approximated by a quadratic function with the error smaller than any quadratic function. More precisely, this means that for some α, β, a, b, c the error

$$f(\mathbf{x}) - f(\mathbf{a}) - [\alpha(\Delta x_1) + \beta(\Delta x_2)] - \frac{1}{2}[a(\Delta x_1)^2 + 2b(\Delta x_1)(\Delta x_2) + c(\Delta x_2)^2]$$

of the quadratic approximation satisfies $\lim_{\mathbf{x}\to\mathbf{a}} \frac{|error(\mathbf{x})|}{|\mathbf{x}-\mathbf{a}|^2} = 0$.

This chain of definitions is easy to continue.

Suppose now that f is differentiable at $\mathbf{a}$ and thus can be approximated by $f(\mathbf{a})+\alpha\Delta x_1+\beta\Delta x_2$ with the required precision. Then the graph of the approximating function is the plane in the (x_1, x_2, y)-space tangent to the graph $y = f(x_1, x_2)$ of the function f at the point $(x_1, x_2, y) = (a_1, a_2, f(a_1, a_2))$. The coefficients α, β are the slopes of the plane in the directions of the coordinate axes x_1, x_2. This shows how to find the coefficients. Restrict the function f to the line $x_2 = a_2$ (respectively $x_1 = a_1$). We get the function $f(x_1, a_2)$ (respectively $f(a_1, x_2)$) in one variable. The derivative of this function at the point $x_1 = a_1$ (respectively $x_2 = a_2$) equals α (respectively β):

$$\alpha = \lim_{\delta\to 0} \frac{f(a_1+\delta, a_2) - f(a_1, a_2)}{\delta}, \quad \beta = \lim_{\delta\to 0} \frac{f(a_1, a_2+\delta) - f(a_1, a_2)}{\delta}.$$

These derivatives are called **partial derivatives** of f and are denoted by $\partial f/\partial x_1$ and $\partial f/\partial x_2$ or, more concisely, f_{x_1} and f_{x_2}. Thus

$$\alpha = f_{x_1}(a_1, a_2) = \frac{\partial f}{\partial x_1}(a_1, a_2), \ \beta = f_{x_2}(a_1, a_2) = \frac{\partial f}{\partial x_2}(a_1, a_2).$$

Examples. (a) The derivative of $e^{\lambda t}$ equals $\lambda e^{\lambda t}$. In fact $f = e^{\lambda t}$ is a function of two variables, t and λ. Computing its time derivative we just think of λ as being constant and therefore compute the partial derivative f_t. The partial derivative $f_\lambda = te^{\lambda t}$.

(b) Let $f(t,x) = e^{-at}\cos\omega x$, $g(t,x) = e^{-at}\sin\omega x$. Then $f_t = -af$, $g_t = -ag$, $f_x = -\omega g$, $g_x = \omega f$. We will come back to this example soon.

(c) Let $f(x_1, x_2) = (ax_1^2 + 2bx_1x_2 + cx_2^2)/2$. Then $f_{x_1} = ax_1 + bx_2$, $f_{x_2} = bx_1 + cx_2$. In particular, $f_{x_1}(0,0) = f_{x_2}(0,0) = 0$.

Functions in the above examples have partial derivatives at any point and therefore the partial derivatives are functions too. We can introduce **2-nd partial derivatives** of a function $y = f(x_1, x_2)$ as partial derivatives of its partial derivatives:

$$\frac{\partial^2 f}{\partial x_1^2} = (f_{x_1})_{x_1}, \ \frac{\partial^2 f}{\partial x_1 \partial x_2} = (f_{x_1})_{x_2}, \ \frac{\partial^2 f}{\partial x_2 \partial x_1} = (f_{x_2})_{x_1}, \ \frac{\partial^2 f}{\partial x_2^2} = (f_{x_2})_{x_2}.$$

Examples. (d) The 2-nd partial derivatives f_{xx}, g_{xx} of the functions from Example (b) are proportional to the functions themselves, $f_{xx} = -\omega g_x = -\omega^2 f$, $g_{xx} = \omega f_x = -\omega^2 g$, and are therefore proportional to their time derivatives f_t and g_t. We will use this fact in our study of the heat equation.

(e) The 2-nd partial derivatives of the quadratic form $f = (ax_1^2+2bx_1x_2+cx_2^2)/2$ are constant:

$$f_{x_1x_1} = a, \quad f_{x_1x_2} = b, \quad f_{x_2x_1} = b, \quad f_{x_2x_2} = c.$$

Suppose now that a function $y = f(\mathbf{x})$ is two times differentiable at the point $\mathbf{a}$ and therefore can be approximated by a quadratic function

$$f(\mathbf{a}) + \alpha\Delta x_1 + \beta\Delta x_2 + [a(\Delta x_1)^2 + 2b(\Delta x_1)(\Delta x_2) + c(\Delta x_2)^2]/2$$

with the required precision. The last example shows how to compute the coefficients a, b, c:

$$a = f_{x_1x_1}(\mathbf{a}), \ b = f_{x_1x_2}(\mathbf{a}) = f_{x_2x_1}(\mathbf{a}), \ c = f_{x_2x_2}(\mathbf{a}).$$

In particular, two of the four 2-nd partial derivatives are equal, $f_{x_1x_2} = f_{x_2x_1}$, provided that the function is two times differentiable.

Warning. The last reservation is not redundant. In fact it is easy to find an example of a function which has all 2-nd partial derivatives at a given point but can not be approximated by a quadratic function with the required precision and hence is not two times differentiable at this point. The mixed partial derivatives of such functions may differ. The same applies to 1-st partial derivatives: the condition that f is differentiable at $\mathbf{a}$ is stronger than the condition that f_{x_1}, f_{x_2} exist at $\mathbf{a}$. However, if all partial derivatives (up to order k) exist in some neighborhood of $\mathbf{a}$ and are continuous at $\mathbf{a}$ then the function is differentiable (k times) at $\mathbf{a}$. In the most of applications we will deal with functions which have all partial derivatives of any order, hence differentiable infinitely many times. Thus the subtle distinction between differentiability and existence of partial derivatives should not scare us.

Exercises 2.2.1.

(a) Compute 1-st and 2-nd partial derivatives of the functions:

$$\sin(x+y), \quad \sin(xy), \quad xy(x-y), \quad \sqrt{x^2+y^2}.$$

(b) Prove that the function $u(t,x) = t^{-1/2}e^{-x^2/4t}$, $t > 0$, satisfies $u_t = u_{xx}$.

(c) Sketch the graph of u as a function of x for small and large $t > 0$. Show that the integral

$$\int_{-\infty}^{\infty} u(t,x)dx$$

does not depend on t.

For those who have studied multivariable calculus: show that the integral equals $\sqrt{4\pi}$.

(d) Prove that a differentiable function $u(x_1, x_2)$ at a point $\mathbf{a} = (a_1, a_2)$ of local maximum or minimum satisfies

$$u_{x_1}(a_1, a_2) = 0 = u_{x_2}(a_1, a_2).$$

2.2.2. Linearization. Consider a time-independent system $\dot{\mathbf{x}} = \mathbf{F}(\mathbf{x})$ of two 1-st order ODEs given by two functions $F_1(x_1, x_2), F_2(x_1, x_2)$.

A point $\mathbf{a}$ on the phase plane of the system is called a singular point (or an equilibrium) if $\mathbf{F}(\mathbf{a}) = \mathbf{0}$. If $\mathbf{a}$ is an equilibrium, then the constant functions $\mathbf{x}(t) = \mathbf{a}$ form a solution to the ODE system, and vice versa.

An equilibrium $\mathbf{a}$ is called asymptotically stable if any solution $\mathbf{x}(t)$ with the initial value $\mathbf{x}(0)$ in a sufficiently small neighborhood of $\mathbf{a}$ approaches $\mathbf{a}$ when t approaches infinity: $\lim_{t\to\infty} \mathbf{x}(t) = \mathbf{a}$. The equilibrium is called asymptotically unstable otherwise, that is if in any neighborhood of $\mathbf{a}$ there exists a point $\mathbf{x}(0)$ whose phase curve does not tend to $\mathbf{a}$.

Examples. (a) The origin $\mathbf{x} = \mathbf{0}$ is the equilibrium for any linear system $\dot{\mathbf{x}} = A\mathbf{x}$. It is asymptotically stable if both roots of the characteristic polynomial have negative real parts (sinks) and is asymptotically unstable otherwise. This should be clear from the phase portraits, or from the fact that eigenvectors $\mathbf{v}$ generate the solutions $e^{\lambda t}\mathbf{v}$ which tend to $\mathbf{0}$ only if the real part of the eigenvalue λ is negative.

(b) In the case of purely imaginary eigenvalues (center) the equilibrium is asymptotically unstable, but it is stable in another, somewhat weaker sense since the phase curves which start in a neighborhood of the origin do not leave this neighborhood. This is the borderline case between spiral sinks and sources.

Let us assume now that the right hand side of the system $\dot{\mathbf{x}} = \mathbf{F}(\mathbf{x})$ is given by functions differentiable as many times as we wish in a neighborhood of an equilibrium $\mathbf{a}$. For simplicity of notations we will assume that $\mathbf{a}$ is the origin on the phase plane. Replacing the functions F_1, F_2 by their approximations $a_{11}x_1 + a_{12}x_2$, $a_{21}x_1 + a_{22}x_2$ near the equilibrium we obtain the system $\dot{\mathbf{x}} = A\mathbf{x}$ of linear ODEs called linearization of the system at $\mathbf{a}$.

Theorem (Stability Criterion).
If all eigenvalues of the linearized system have negative real parts then the equilibrium of the non-linear system is asymptotically stable.

Remark. In fact if at least one of the eigenvalues has positive real part then the equilibrium of the non-linear system is asymptotically unstable (and is not stable even in the sense mentioned in Example (b)). If however the eigenvalues are on the imaginary axis, the linearization does not allow to judge about stability in the non-linear system (see Exercises).

Sketch of the proof. The idea is to show that in some coordinate system (y_1, y_2) near the equilibrium all phase curves of the non-linear system cross the concentric circles $y_1^2 + y_2^2 = const$ inward and to deduce from this that the phase curves inevitably approach the origin $\mathbf{y} = \mathbf{0}$ when $t \to \infty$. Although we can not compute the phase curves (and do not even specify the functions F_1, F_2) we know that the velocity vectors of the phase curves are well approximated by those of the linearized system in a vicinity of the origin. Thus if the phase curves of the linearized system cross the circles inward the same will be true for the non-linear system. According to the similarity classification of matrices given in 1.5.3 the linear system $\dot{\mathbf{x}} = A\mathbf{x}$ after a suitable change of coordinates takes on one of the three normal forms described 2.1.2. If the eigenvalues have negative real parts, the linear system is a spiral, nodal or degenerate sink. We can examine the three standard ODE systems and their solutions in order to figure out whether the phase curves have the desired property.

In fact the property to cross the circles $y_1^2 + y_2^2 = const$ inward is obvious in the case of the spiral sink since the motion of the phase points is described in this case as a simultaneous rotation and contraction on the (y_1, y_2)-plane. In the case of the nodal sink $\dot{y}_1 = \lambda y_1$, $\dot{y}_2 = \lambda y_2$ the property is also easy to check since the squared distance to the origin $y_1(t)^2 + y_2(t)^2 = y_1^2(0)e^{2\lambda_1 t} + y_2^2(0)e^{2\lambda_2 t}$ decreases monotonously with time when both λ_1, λ_2 are negative. In the case of the Jordan cell $\dot{y}_1 = \lambda y_1 + y_2$, $\dot{y}_2 = \lambda y_2$ we first rescale the coordinate y_2 to εy_2 in order to bring the system to the form $\dot{y}_1 = \lambda y_1 + \varepsilon y_2$, $\dot{y}_2 = \lambda y_2$. Now the time derivative of $y_1^2(t) + y_2^2(t)$ equals

$$2y_1\dot{y}_1 + 2y_2\dot{y}_2 = 2y_1(\lambda y_1 + \varepsilon y_2) + 2\lambda y_2^2 = 2\lambda y_1^2 + 2\varepsilon y_1 y_2 + 2\lambda y_2^2.$$

This is a quadratic form with the matrix $\begin{bmatrix} 2\lambda & \varepsilon \\ \varepsilon & 2\lambda \end{bmatrix}$. The determinant $4\lambda^2 - \varepsilon^2$ of the matrix is positive if ε has been chosen smaller than $|2\lambda|$. Since λ is negative, the quadratic form is negative for any $\mathbf{y} \neq \mathbf{0}$ according to our classification of quadratic forms and Example 1.5.2(k). This means that the distance to the origin decreases along the phase curves in this case too. □

In the next section we will try the stability criterion in action in the study of an ecological model.

Exercises 2.2.2.

(a) Is the lower (upper) equilibrium of the undamped pendulum asymptotically stable?

(b) Find out how asymptotical stability of the upper and lower equilibria of the damped pendulum depends on the damping coefficient k. Sketch the phase portraits of the non-linear system on the phase plane with coordinates $(x, \dot{x})$.

(c) Consider the ODE $\ddot{x} = -k\dot{x}^3 - \sin x$ of the pendulum in the presence of damping depending cubically on the angular velocity. Show that linearization of the equation near the equilibrium $(x, \dot{x}) = (0, 0)$ has imaginary eigenvalues.

(d) Let $x(t)$ be a non-constant solution to the ODE of Problem (c). Show that the energy

$$E(t) = \frac{\dot{x}^2(t)}{2} - \cos x(t)$$

is a decreasing function of t if $k > 0$ and an increasing function of t if $k < 0$. Deduce from this that the equilibrium $(x, \dot{x}) = (0, 0)$ is asymptotically stable when $k > 0$ and asymptotically unstable when $k < 0$.

2.2.3. Competing species. According to Example 2.1.1(b) the equations $\dot{x}_1 = \lambda_1 x_1$, $\dot{x}_2 = \lambda_2 x_2$ describe reproduction of two species provided that the food supply is unlimited. If however the species have to compete for food with each other and with their own kind, the reproduction coefficients λ_1, λ_2 will no longer remain constant, but will depend on the amounts x_1, x_2. Accepting the simplest model where the reproduction coefficients decrease linearly when x_1 and x_2 grow, we arrive at the system of ODEs

$$\begin{array}{rcl} \dot{x}_1 & = & x_1(k_1 - a_1 x_1 - b_1 x_2) \\ \dot{x}_2 & = & x_2(k_2 - a_2 x_1 - b_2 x_2) \end{array}.$$

We assume that the reproduction rates k_1, k_2 and the competition factors a_1, b_1, a_2, b_2 are all positive.

The system has four equilibria (which are found by equating the right hand sides to zero).

- $x_1 = x_2 = 0$. It is the origin on the phase plane. Linearizing the system at this point we find $\dot{x}_1 = k_1 x_1$, $\dot{x}_2 = k_2 x_2$ which is a nodal source. Thus the equilibrium is unstable. This means that when the amounts x_1, x_2 of the species are small compared to the food supply, the species do not really compete and both populations grow exponentially.
- $x_1 = 0, x_2 = k_2/b_2$. It is the intersection point of the line $a_2 x_1 + b_2 x_2 = k_2$ (we denote it L_2) with the x_2-axis. The first species is extinct, the second one achieves an equilibrium with the food supply being just enough in order to keep the population.
- $x_2 = 0, x_1 = k_1/a_1$. It is the intersection point of the line $L_1 : a_1 x_1 + b_1 x_2 = k_2$ with the x_1-axis. The second species is extinct, the first one is in the equilibrium with the food supply.

  The stability question about these two equilibria is whether a small amount of the other species injected to the system will extinct.
- (x_1, x_2) is the intersection point of the lines L_1 and L_2. This equilibrium makes "ecological" sense only if $x_1, x_2 > 0$, that is if the lines intersect in the 1-st quadrant of the phase plane. If this happens, the equilibrium corresponds to coexistence of both species in the system. However, such coexistence of species can not be achieved in the real system unless the equilibrium is stable. (In fact in our model, if this equilibrium is unstable, small random deviations from it — which inevitably occur in real systems — will eventually drive one of the species to extinction.)

The result of stability analysis of the four equilibria in the system turns out to depend on the mutual position of the lines L_1, L_2 in the 1-st quadrant and can be described graphically as it is shown on the picture. In the case I (L_1 is above L_2) the coexistence equilibrium is not present, and only the equilibrium with extinct second species is stable. In the case II (L_2 is above L_1) the situation is the same, but the species switch their roles. In the case III the coexistence equilibrium is unstable, and the other two are both stable. Thus, only one of the species survives, but which one — depends on the initial amounts $x_1(0), x_2(0)$. In the case IV the coexistence equilibrium is the only stable one. This case is singled out by the inequality $a_1/a_2 > k_1/k_2 > b_1/b_2$. We can interpret the result this way: coexistence of species in our model is possible only if competition of both species with their own kind is tougher than competition with the other species.

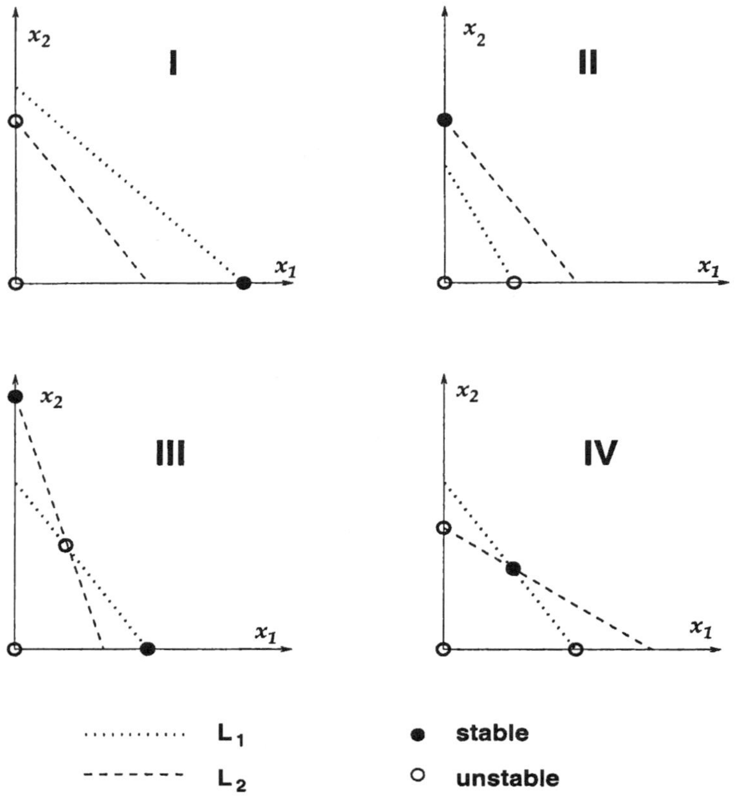

We will carry out the stability analysis for the coexistence equilibrium and leave the analysis of the remaining two equilibria as an exercise.

In order to simplify computations it is convenient to rescale the variables x_1, x_2 and t in such a way that the ODE system assumes the form

$$\begin{array}{rcl} dX_1/d\tau & = & X_1(k - aX_1 - bX_2) \\ dX_2/d\tau & = & X_2(1 - X_1 - X_2) \end{array},$$

where $\tau = k_2 t$, $X_1 = a_2 x_1/k_2$, $X_2 = b_2 x_2/k_2$ and $k = k_1/k_2$, $a = a_1/a_2$, $b = b_1/b_2$. The intersection point of the lines L_1, L_2 has coordinates $A_1 = (k-b)/(a-b)$, $A_2 = (k-a)/(b-a)$. The intersection point is in the 1-st quadrant only if $k-b, a-k$ and $a-b$ have the same sign. Partial derivatives of the functions $F = X_1(k-aX_1-bX_2)$ and $G = X_2(1 - X_1 - X_2)$ *evaluated at the intersection point* form the matrix of the linearized ODE system:

$$A = \begin{bmatrix} F_{X_1}(A_1, A_2) & F_{X_2}(A_1, A_2) \\ G_{X_1}(A_1, A_2) & G_{X_2}(A_1, A_2) \end{bmatrix} = \begin{bmatrix} -aA_1 & -bA_1 \\ -A_2 & -A_2 \end{bmatrix}.$$

The discriminant of the characteristic equation equals

$$(\operatorname{tr} A)^2 - 4\det A = (aA_1 + A_2)^2 - 4(a-b)A_1A_2 = (aA_1 - A_2)^2 + 4bA_1A_2$$

and is positive under our hypothesis that the coordinates A_1, A_2 of the equilibrium are positive. The roots of the characteristic equation are real and given by the formula

$$\frac{-(aA_1 + A_2) \pm \sqrt{(aA_1 + A_2)^2 - 4(a-b)A_1A_2}}{2}.$$

They are both negative if and only if $a - b > 0$. Thus the coexistence equilibrium is both in the 1-st quadrant and asymptotically stable when $k - b > 0$ and $a - k > 0$, that is when $a > k > b$.

Exercises 2.2.3.

(a) Linearize the competing species system near the equilibria $(x_1, x_2) = (0, k_2/b_2)$ and $(x_1, x_2) = (k_1/a_1, 0)$, and examine stability of the equilibria.

(b) Study linearizations of all equilibria in the cases I – IV in order to find their place in the classification and sketch their phase portraits. Pay attention to positions of eigenlines.

(c) Sketch plausible phase portraits of the non-linear system in the cases I – IV in the 1-st quadrant of the phase plane.

2.3. PDE

Partial Differential Equations are called so because they relate partial derivatives of unknown functions. We will study here only one example, namely the PDE describing the heat conduction phenomenon. It also can be considered as a dynamical system, but in contrast with ODEs the phase space here has infinite dimension.

2.3.1. The heat equation. In the beginning of XIX-th century, when Joseph Fourier introduced the heat equation, heat was thought of as some invisible fluid that can leak from one part of a body to another. Although the microscopic nature of heat turned out to be different, the heat equation remains an adequate mathematical model of the phenomenon because it is derived from very basic macroscopic hypotheses.

Let $u(t,x,y,z)$ denote the temperature at the moment t at the point with coordinates x,y,z in a 3-dimensional homogeneous medium. Consider the temperature $u(t+\Delta t,x,y,z)$ at the same point at a moment close to t. The temperature increment $\Delta u = u(t+\Delta t,x) - u(t,x)$ is caused by some additional amount of heat ΔH accumulated in the volume $\Delta x\Delta y\Delta z$ near the point. The amount ΔH is proportional to the volume and to the temperature increment Δu with the proportionality coefficient c characterizing the material and called thermal capacity. This proportionality law agrees well with experimental data if temperature variations are not too large. Thus

$$\frac{\Delta H}{\Delta t\Delta x\Delta y\Delta z} \approx c\frac{\Delta u}{\Delta t}.$$

The additional amount of heat ΔH received by the volume $\Delta x\Delta y\Delta z$ during the time interval Δt is caused by the heat flow through the walls of the volume. One assumes that the flow across the wall $x = const$ from left to right is proportional to the time interval Δt, to the surface area $\Delta y\Delta z$ and to *minus* the slope $\partial u/\partial x$ of the temperature profile in the x-direction. This proportionality law is the simplest hypothesis which agrees with our experience that heat leaks from warmer parts to cooler parts of the body and that the heat flow vanishes if the parts have the same temperature. The proportionality coefficient $k > 0$ called thermal conductivity is another characteristic of the material.

For the sake of simplicity we will assume further on that the temperature is constant along each plane $x = const$ and thus depends only on t and x. In particular the derivatives u_y and u_z and respectively the heat flows in y- and z-directions vanish. We conclude that the amount of heat ΔH is due to the difference of the heat flow across the walls x and $x + \Delta x$:

$$\frac{\Delta H}{\Delta t\Delta x\Delta y\Delta z} \approx k\frac{u_x(t,x+\Delta x) - u_x(t,x)}{\Delta x}.$$

Equating the two expressions for ΔH and passing to the limit $\Delta t \to 0$, $\Delta x \to 0$ we obtain the heat equation

$$c\frac{\partial u}{\partial t}(t,x) = k\frac{\partial^2 u}{\partial x^2}(t,x).$$

It can be rewritten as

$$u_t = \alpha^2 u_{xx}, \text{ where } \alpha^2 = k/c > 0$$

is the ratio of thermal conductivity and capacity called thermal diffusivity coefficient.

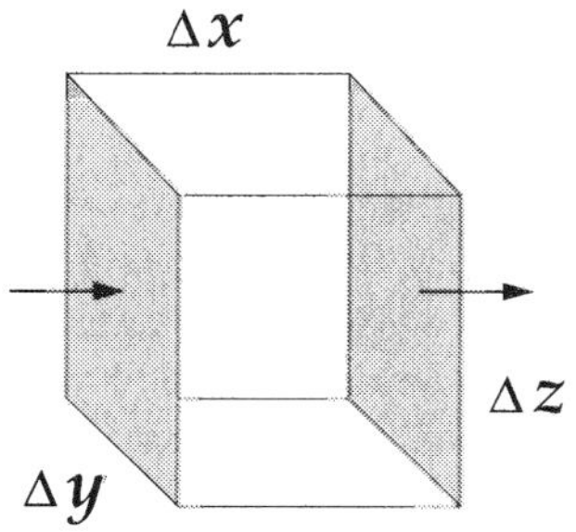

Exercises 2.3.1.

(a) Suppose that the bar is not homogeneous so that the thermal capacity c and the thermal conductivity k depend on x. Show that the heat conduction in the non-homogeneous bar is described by the equation

$$c(x)u_t = (k(x)u_x)_x.$$

(b) Assuming that the temperature $u(t, x, y, z)$ in a homogeneous medium may depend on all the 3 space coordinates derive the $3D$ heat equation

$$u_t = \alpha^2(u_{xx} + u_{yy} + u_{zz}).$$

(c) A drop of ink in a glass of water eventually "dissolves" (even without mixing) due to the **diffusion** phenomenon. (It is caused by random microscopic motion of ink particles called **Brownian motion**). Let $u(t, x, y, z)$ denote concentration of ink in water. Let us assume that the diffusion flow is proportional to minus the gradient of concentration. (The proportionality coefficient α^2 called the **diffusion coefficient** of ink in water.) Show that the concentration function satisfies the **diffusion equation**

$$u_t = \alpha^2(u_{xx} + u_{yy} + u_{zz})$$

identical to the heat equation.

(d) Suppose that at the moment $t = 0$ the ink drop of mass m is concentrated at the point $x = 0$ of an infinite 1-dimensional "glass" (tube). Assuming that the diffusion coefficient of ink in water equals 1, show that the mass distribution at the moment $t > 0$ is described by the formula

$$\frac{m}{\sqrt{4\pi t}} e^{-x^2/4t}.$$

Remark. In is tempting in a mathematical study of the heat equation with one space variable to talk about heat conduction in a homogeneous bar of length L with thermal diffusivity α^2. We are not going to resist this temptation and will assume that the left end of the bar is positioned at $x = 0$ and the right end — at $x = L$. It is worth mentioning however that the heat equation $u_t = \alpha^2 u_{xx}$ applies to a real bar only if the side surface of the bar is well insulated against heat leakage. In real life this condition is seldom satisfied. A real phenomenon described well by the heat equation is heat conduction in a $3D$-layer of width L to be small compared to the y- and z-dimensions of the layer. In this case the heat leakage through the side surface will alter the temperature in the interior of the layer very little for the surface is small compared to the area of the faces $x = 0$ and $x = L$.

2.3.2. Boundary value problems. The heat equation is to describe time evolution of temperature distribution in a solid homogeneous bar. We may expect that the the process depends on the initial temperature distribution $u(0, x)$. However it is easy to see that the initial distribution alone is not sufficient in order to determine the solution unambiguously since the process also depends on the conditions maintained at the extremities of the bar.

Example. (a) A time-independent function satisfies the heat equation if and only if it is a polynomial of degree ≤ 1: $u(t, x) = ax + b$. What is the physical meaning of such solutions? If the bar is completely insulated, and the initial temperature along the bar is constant then no heat flow will ever arise and the temperature will remain constant. However our experience suggests that if the initial temperature is not constant, then it will evolve toward the constant distribution. In particular, the function $u = ax + b$ with $a \neq 0$, although satisfies the heat equation, does not describe the process correctly. In fact this solution describes a steady heat flow $-ku_x = -ka$ along the bar and assumes that the heat is supplied at one end of the bar and carried away at the other end. In particular, the bar is not insulated, but instead the temperatures at the ends are maintained constant and equal to $T_0 = b$ at $x = 0$ and $T_L = aL + b$ at $x = L$.

We will consider two physically meaningful types of **boundary conditions**: (i) **no heat flow** through the boundaries (which means that the bar is completely insulated) and (ii) given **constant temperatures** to be maintained at the ends of the bar. The boundary conditions are expressed mathematically as

$$(i)\;\; u_x(t, 0) = 0 = u_x(t, L) \;\;\text{and}\;\; (ii)\; u(t, 0) = T_0,\; u(t, L) = T_L$$

for all $t > 0$.

Example. (b) Suppose that we want to find temperature evolution $u(t, x)$ in a bar with the ends maintained at the temperatures T_0 and T_L provided that the initial temperature distribution $u(0, x) = \phi(x)$ is given. The function $T_0 + (T_L - T_0)x/L$ satisfies the heat equation and the boundary conditions. Put $v(t, x) = u(t, x) - T_0 - (T_L - T_0)x/L$. If u satisfies the heat equation $u_t = \alpha^2 u_{xx}$ then v satisfies the same equation. It satisfies the **zero temperature** boundary conditions $v(t, 0) = 0 = v(t, L)$ and the initial condition $v(0, x) = \psi(x)$ with $\psi(x) = \phi(x) - T_0 - (T_L - T_0)x/L$. Thus if we knew how to solve the heat equation for arbitrary initial temperature distributions and zero temperature boundary conditions we would be able to find the solution with constant temperature boundary conditions.

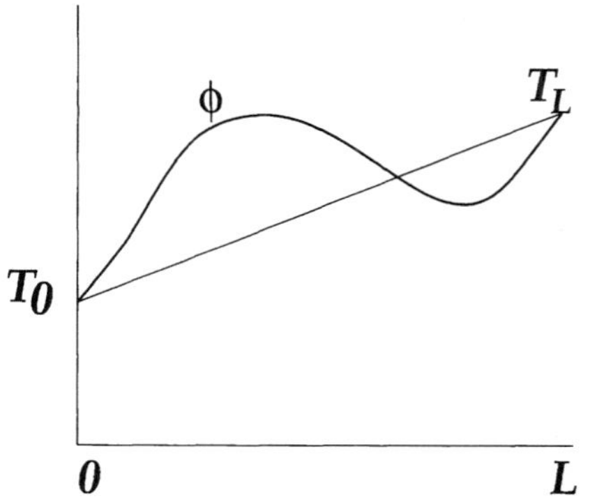

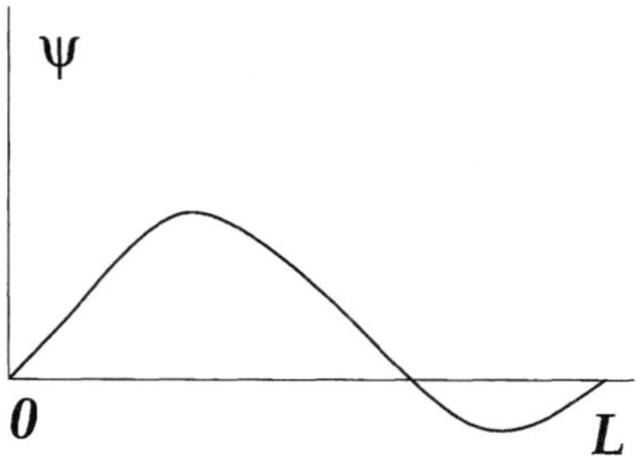

Exercises 2.3.2.

(a) A homogeneous solid bar of length L has zero temperature at $t = 0$, and for $t > 0$ the extremities of the bar are maintained at the constant temperature T_0. Write down the initial/boundary value problem for the temperature function u and reduce it to the problem with zero-temperature boundary conditions.

(b) The initial temperature distribution in a homogeneous solid bar of length π and of thermal diffusivity $\alpha^2 = 1$ is given by the function $u(0, x) = \sin^2 x$ for $0 < x < \pi$, and the extremities of the bar are completely insulated. Find the temperature distribution $u(t, x)$.

(c) The initial temperature distribution in the bar of length π and of thermal diffusivity α^2 is $u(0, x) = \sin^3 x$ for $0 < x < \pi$, and the extremities of the bar are maintained at zero temperatures. Find the temperature distribution $u(t, x)$.

(d) The temperature of the left end $x = 0$ of the bar is maintained at the zero level, and the right end $x = L$ is insulated. Find the temperature distribution $u(t, x)$ in the bar if the initial temperature distribution equals

$$\sin\frac{\pi x}{2L}, \ \sin\frac{3\pi x}{2L}, \ \sin\frac{5\pi x}{2L}, \ \ldots$$

(e) Let $u(t, x)$ be a solution to the heat equation $u_t = u_{xx}$ in the region $0 \le x \le \pi$, $0 < t < \infty$ satisfying the zero-temperature (no-flow) boundary condition. Show that

$$v(t, x) = u(\frac{\alpha^2\pi^2 t}{L^2}, \frac{\pi x}{L})$$

satisfies the heat equation $v_t = \alpha^2 v_{xx}$ in the region $0 \le x \le L$, $0 < t < \infty$ and the corresponding boundary conditions.

We may therefore concentrate our efforts on the following two initial / boundary value problems for the heat equation:

- find a function $u(t,x)$ satisfying the heat equation $u_t = \alpha^2 u_{xx}$ in the semi-infinite strip $t \geq 0,\ 0 \leq x \leq L$, the no–flow boundary conditions $u_x(t,0) = 0 = u_x(t,L)$ for $t > 0$ and the initial condition $u(0,x) = \phi(x)$ for $0 < x < L$;
- find a function $v(t,x)$ satisfying the heat equation $v_t = \alpha^2 v_{xx}$ for $t \geq 0,\ 0 \leq x \leq L$, the zero–temperature boundary conditions $v(t,0) = 0 = v(t,L)$ for $t > 0$ and the initial condition $v(0,x) = \psi(x)$.

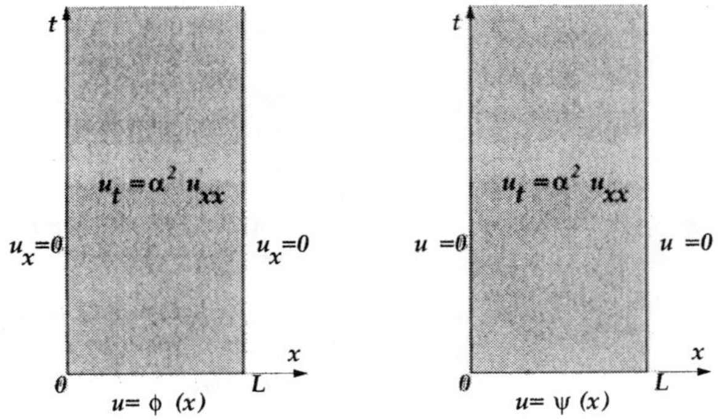

Examples. (c) The function $u(t,x) = e^{-at}\cos\omega x$ satisfies the equation $u_t = \alpha^2 u_{xx}$ if $a = \alpha^2\omega^2$. It satisfies the boundary condition $u_x(t,L) = 0$ if $\omega = \pi n/L,\ n = 0,1,2,...$ (and $u_x(t,0) = 0$ for any ω). Thus the functions

$$u_n(t,x) = e^{-\alpha^2\pi^2 n^2 t/L^2}\cos\frac{\pi n x}{L},\ n = 0,1,2,...,$$

are particular solutions to the no–flow boundary value problem with the initial condition $\phi(x) = \cos\frac{\pi n x}{L}$.

(d) Similarly, the functions

$$v_n(t,x) = e^{-\alpha^2\pi^2 n^2 t/L^2}\sin\frac{\pi n x}{L},\ n = 1,2,...,$$

are particular solution to the zero–temperature boundary value problem with the initial condition $\psi(x) = \sin\frac{\pi n x}{L}$.

The function $au(t,x) + bv(t,x) + ... + cw(t,x)$ is called a **linear combination** of the functions $u, v, ..., w$.

The heat equation is linear: linear combinations of solutions are solutions too. The no–flow (zero–temperature) boundary condition is linear: linear combinations of functions satisfying the boundary condition satisfy the same boundary condition. Initial conditions of such linear combinations of solutions are the linear combinations of the initial conditions of those solutions. Applying this linearity principle to the particular solutions u_n we obtain solutions $u = a_0u_0 + a_1u_1 + ... + a_nu_n$ (and similarly, $v = b_1v_1 + ... + b_nv_n$) to our boundary value problems with the initial conditions given by trigonometric polynomials

$$\begin{array}{lll} \phi(x) & = a_0 + & a_1\cos\frac{\pi x}{L} + ... + a_n\cos\frac{\pi n x}{L} \\ \psi(x) & = & b_1\sin\frac{\pi x}{L} + ... + b_n\sin\frac{\pi n x}{L} \end{array}.$$

We will see that such solutions are sufficient in order to approximate any other solutions to our problems as precisely as necessary.

2.4. Fourier series

A general approach to the initial/boundary value problems for the heat equation is based on representation of periodic functions by infinite linear combinations of trigonometric functions known as Fourier series.

2.4.1. Fourier coefficients. A function $y = f(x)$ is called $2L$-periodic if $f(x+2L) = f(x)$ for any x. In the theory of Fourier series it is convenient to consider periodic functions (of the real variable x) which assume complex values: $f(x) = u(x) + iv(x)$, where the real and imaginary parts are real-valued $2L$-periodic functions.

Example. (a) $e^{\pi inx/L} = \cos\frac{\pi nx}{L} + i\sin\frac{\pi nx}{L}$, $n = 0, \pm1, \pm2, ...$, are complex-valued $2L$-periodic functions.

The inner product of two $2L$-periodic functions is defined as

$$\langle f, g\rangle = \frac{1}{2L}\int_{-L}^{L} f(x)\overline{g(x)}\,dx.$$

Here $\bar{g}$ is the function complex conjugate to g. The integral of a complex-valued function is defined by integration of its real and imaginary parts and is generally speaking a complex number. It is easy to see that

$$\langle g, f\rangle = \overline{\langle f, g\rangle} \text{ and } \langle \lambda f + \mu g, h\rangle = \lambda\langle f, h\rangle + \mu\langle g, h\rangle$$

for any complex numbers λ, μ. Two periodic functions are called orthogonal if $\langle f, g\rangle = 0$.

Example. (b) The functions $e^{\pi inx/L}, n = 0, \pm1, \pm2, ...$, are pairwise orthogonal: if $m \neq n$

$$\langle e^{\pi imx/L}, e^{\pi inx/L}\rangle = \frac{1}{2L}\int_{-L}^{L} e^{\pi i(m-n)x/L}dx = \frac{1}{2\pi i(m-n)}e^{\pi i(m-n)x/L}|_{-L}^{L} = 0.$$

For $m = n$ we have $\langle e^{\pi inx/L}, e^{\pi in/L}\rangle = 1$.

A series of the form $\sum_{m=-\infty}^{\infty} c_m e^{\pi imx/L}$ is a $2L$-periodic complex Fourier series with complex coefficients c_m, $m = 0, \pm1, \pm2, \ldots$. Suppose that such a series converges to some function $f(x)$. Computing the inner products

$$\langle f, e^{\pi inx/L}\rangle = \sum_m c_m\langle e^{\pi imx/L}, e^{\pi inx/L}\rangle = c_n$$

(where we assume that integration of the infinite sum can be replaced by summation of the term-wise integrals) we find the following formula for the Fourier coefficients of the function f:

$$c_n = \frac{1}{2L}\int_{-L}^{L} f(x)e^{-\pi inx/L}dx.$$

Example. (c) Let ψ be a $2L$-periodic function equal x for $-L < x < L$. We have $c_0 = \frac{1}{2L}\int_{-L}^{L} xdx = 0$. Integrating by parts we find for $n \neq 0$

$$c_n = \frac{1}{2L}\int_{-L}^{L} xe^{-\pi inx/L}dx = \frac{-x}{2\pi in}e^{-\pi inx/L}|_{-L}^{L} + \frac{1}{2\pi in}\int_{-L}^{L} e^{-\pi inx/L}dx =$$

$$\frac{-L}{2\pi in}(-1)^n - \frac{L}{2\pi in}(-1)^n + 0 = \frac{iL(-1)^n}{\pi n}.$$

Exercises 2.4.1.
(a) Show that the inner product of *real* $2L$-periodic functions is symmetric and bilinear.
(b) Prove that the $2L$-periodic functions

$$\cos\frac{\pi n x}{L}, n = 0, 1, 2, ..., \quad \sin\frac{\pi n x}{L}, n = 1, 2, 3, ...,$$

are pairwise orthogonal and find their inner squares.
(c) Let $f = \sum c_n e^{\pi i n x/L}$ be a complex Fourier series. Derive Parseval's identity:

$$\langle f, f\rangle = \sum |c_n|^2.$$

Apply Parseval's identity to the function ψ of Example 2.4.1(c) in order to show that

$$1 + \frac{1}{4} + \frac{1}{9} + ... + \frac{1}{n^2} + ... = \frac{\pi^2}{6}.$$

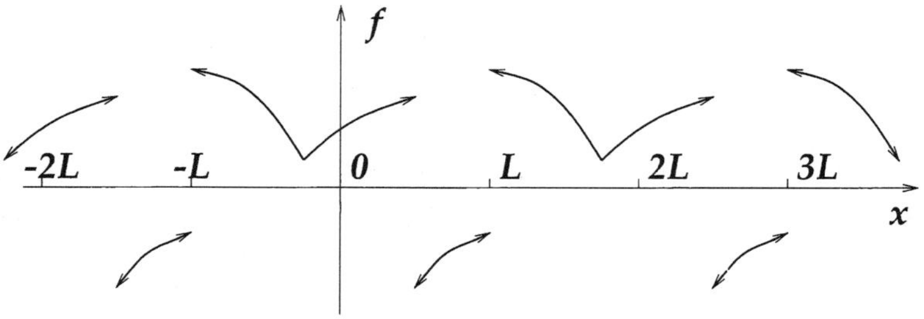

A piecewise differentiable 2L-periodic function

2.4.2. Convergence. Piecewise continuity of f is sufficient in order to define the Fourier coefficients. It is not sufficient however for convergence of the Fourier series to the function f. We introduce a class of functions convenient for our applications, namely piecewise differentiable periodic functions. By definition this means that on the periodicity interval $[-L, L]$ (i) the function f is differentiable (and therefore continuous) everywhere except may be finitely many points, (ii) at each discontinuity point x it has finite right and left limits $f(x^+)$ and $f(x^-)$, and (iii) the derivative f' is continuous everywhere except may be finitely many points and has right and left limits at each discontinuity point. Notice that the actual values of the function at finitely many discontinuity points can not affect the integrals defining the Fourier coefficients and therefore may have nothing to do with the sum of the Fourier series at such points.

Theorem (Fourier Convergence Theorem).
At each point x the Fourier series of a piecewise differentiable $2L$-periodic function f converges to the mean $(f(x^+) + f(x^-))/2$ of the right and left limits of f (and therefore converges to $f(x)$ if f is continuous at x).

Example. The function ψ of Example 2.4.1(c) has the Fourier series

$$\frac{iL}{\pi} \sum_{n\neq 0} \frac{(-1)^n}{n} e^{\pi i n x/L}.$$

The values of the n-th term at $x = 0$ and $x = L$ cancel with the values of the $-n$-th term at the corresponding point. Thus the series has zero sum at $x = 0, L$. The function ψ is continuous at $x = 0$ with $\psi(0) = 0$, and is discontinuous at $x = L$ with the limits $\psi(L^+) = -1$, $\psi(L^-) = 1$. The values agree with the statement of the Fourier theorem. At $x = L/2$ the terms $\frac{(-1)^n}{n} e^{\pi i n/2} = (-i)^n/n$ of the Fourier series with even $n = \pm 2k$ cancel each other, and the terms with odd $n = \pm(2k+1)$ yield

$$\frac{2L}{\pi} \sum_{k=0}^{\infty} \frac{(-1)^k}{(2k+1)}.$$

According to the Fourier theorem the series converges to $\psi(L/2) = L/2$. Thus we deduce *from the theorem* that

$$1 - \frac{1}{3} + \frac{1}{5} - \ldots + \frac{(-1)^k}{(2k+1)} + \ldots = \frac{\pi}{4}.$$

The last identity is quite non-trivial and illustrates the power of the Fourier convergence theorem. It is not surprising therefore that the proof of the theorem is non-trivial too. Fourier himself, who tried to find a proof for many years, did not really succeed, and we will not even try to present the prove here.

Exercises 2.4.2. (a) Compute complex Fourier coefficients of the $2L$-periodic function equal 1 for $0 < x < L$, -1 for $-L < x < 0$ and 0 for $x = 0, \pm L$. Apply Parseval's identity to this function in order to compute $\sum 1/(2k-1)^2$. Using Fourier Convergence Theorem show that this function equals the sum of the series

$$\frac{4}{\pi} \sum_{k=0}^{\infty} \frac{1}{2k-1} \sin \frac{\pi n x}{L}.$$

Graph a few first terms of the series and their sum in order to see how the function is approximated by the series. Check the statement of the theorem at $x = L/2$.

(b) Are the (non-periodic) functions $x^{-1/2}$, $x^{1/2}$, $x^{3/2}$, $\sin(1/x)$, $x\sin(1/x)$, $x^2\sin(1/x)$, $x^3\sin(1/x)$ piecewise differentiable?

2.4.3. Real even and odd functions. If the $2L$-periodic function f is real, that is $\overline{f(x)} = f(x)$, then the Fourier coefficients c_n and c_{-n} are complex conjugate. Put $c_n = (a_n - ib_n)/2$ for $n \geq 0$ (note that $b_0 = 0$ since c_0 is real). Using Euler's formula $e^{-\pi i n x/L} = \cos\frac{\pi n x}{L} - i \sin\frac{\pi i n x}{L}$ we find

$$a_n = \frac{1}{L} \int_{-L}^{L} f(x) \cos \frac{\pi n x}{L}\, dx, \ n = 0, 1, 2, \ldots$$

$$b_n = \frac{1}{L} \int_{-L}^{L} f(x) \sin \frac{\pi n x}{L}\, dx \ , n = 1, 2, \ldots$$

On the other hand, the sum of the $\pm n$-th terms of the Fourier series can be rewritten as

$$a_n \frac{e^{\pi i n x/L} + e^{-\pi i n x/L}}{2} + b_n \frac{e^{\pi i n x/L} - e^{-\pi i n x/L}}{2i} = a_n \cos \frac{\pi n x}{L} + b_n \sin \frac{\pi n x}{L}.$$

Thus, a piecewise differentiable real function f is represented by its real Fourier series

$$\frac{a_0}{2} + \sum_{n=1}^{\infty} a_n \cos \frac{\pi n x}{L} + \sum_{n=1}^{\infty} b_n \sin \frac{\pi n x}{L}.$$

A function $y = f(x)$ is called even if $f(-x) = f(x)$ for all x and is called odd if $f(-x) = -f(x)$ for all x. Graphs of even functions are symmetric about the y-axis, and graphs of odd functions are centrally symmetric about the origin. The functions $\cos\frac{\pi n x}{L}$ are even, while $\sin\frac{\pi n x}{L}$ are odd.

Let f be a real $2L$-periodic even function. Then the Fourier coefficients b_n vanish (why?) and only even terms in the Fourier series survive. We obtain the cosine Fourier series representing piecewise differentiable even functions, and a formula for their Fourier coefficients which takes in account the symmetry property:

$$f(x) = \frac{a_0}{2} + \sum_{n=1}^{\infty} a_n \cos\frac{\pi n x}{L}, \quad a_n = \frac{2}{L}\int_0^L f(x)\cos\frac{\pi n x}{L}dx.$$

Similarly, real odd $2L$-periodic piecewise differentiable functions are represented by the sine Fourier series

$$f(x) = \sum_{n=1}^{\infty} b_n \sin\frac{\pi n x}{L}, \quad b_n = \frac{2}{L}\int_0^L f(x)\sin\frac{\pi n x}{L}dx.$$

Example. (a) The function ψ from Examples 2.4.1(c) is real and odd. Combining the symmetric terms of its complex Fourier series we find for $n > 0$

$$\frac{(-1)^n iL}{\pi n}[e^{\pi i n x/L} - e^{-\pi i n x/L}] = (-1)^{n-1}\frac{2L}{\pi n}\sin\frac{\pi n x}{L}.$$

Thus we have found the sine Fourier series of the function:

$$\psi(x) = \frac{2L}{\pi}\sum_{n=1}^{\infty}\frac{(-1)^{n-1}}{n}\sin\frac{\pi n x}{L}.$$

(b) Let ϕ be the $2L$-periodic function equal to $|x|$ on the interval $[-L, L]$. It is real, even, piecewise differentiable, has no discontinuity points and is therefore equal to the sum of its cosine Fourier series. We find $a_0 = \frac{2}{L}\int_0^L x dx = L$, and for $n > 0$ integration by parts yields:

$$a_n = \frac{2}{L}\int_0^L x\cos\frac{\pi n x}{L}dx = \frac{2x}{\pi n}\sin\frac{\pi n x}{L}\,|_0^L - \frac{2}{\pi n}\int_0^L \sin\frac{\pi n x}{L}dx =$$

$$0 - 0 + \frac{2L}{\pi^2 n^2}\cos\frac{\pi n x}{L}\,|_0^L = \frac{2L}{\pi^2 n^2}[(-1)^n - 1].$$

Thus

$$\phi(x) = \frac{L}{2} - \frac{4L}{\pi^2}\sum_{k=1}^{\infty}\frac{1}{(2k-1)^2}\cos\frac{\pi(2k-1)x}{L}.$$

In particular, substituting $x = 0$ we derive from the Fourier theorem that

$$1 + \frac{1}{9} + \frac{1}{25} + \dots + \frac{1}{(2k-1)^2} + \dots = \frac{\pi^2}{8}.$$

Exercises 2.4.3. Represent the $2L$-periodic functions f, g, h by real Fourier series:

$$(a)\ \ f(x) = \begin{cases} 1 \text{ for } |x| < L/2 \\ 0 \text{ for } L/2 < |x| < L \end{cases},$$

$$(b)\ \ g(x) = \begin{cases} \text{sign}(x) \text{ for } 0 < |x| < L/2 \\ 0 \text{ for } L/2 < |x| < L \end{cases},$$

$$(c)\ \ h(x) = x^2 \text{ for } -L < x < L\ .$$

2.5. The Fourier method

In the application of Fourier series to the heat equation, the idea is to represent the initial function on the interval $(0, L)$ by a $2L$-periodic sine or cosine Fourier series (depending on the type of the boundary conditions) and then use our knowledge of particular solutions with trigonometric initial values.

2.5.1. The series solution. In the zero–temperature boundary value problem we are looking for a solution $v(t, x)$ to the heat equation $v_t = \alpha^2 v_{xx}$ in the strip $t \geq 0,\ 0 \leq x \leq L$ satisfying a given initial condition $v(0, x) = \psi(x),\ 0 < x < L$, and the boundary conditions $v(t, 0) = 0 = v(t, L),\ t > 0$. The initial value function ψ defined on the interval $(0, L)$ can be extended to an *odd* function on the interval $(-L, L)$ and then extended to the entire line as a $2L$-periodic odd function. Assuming that ψ is piecewise differentiable we can expand it into the *sine* Fourier series with coefficients

$$b_n = \frac{2}{L} \int_0^L \psi(x) \sin \frac{\pi n x}{L} dx.$$

Then the series

$$v(t, x) = \sum_{n=1}^{\infty} b_n e^{-\alpha^2 \pi^2 n^2 t / L^2} \sin \frac{\pi n x}{L}$$

(i) satisfies the heat equation term-wise, (ii) at $t = 0$ converges to the function $\psi(x)$ on the interval $(0, L)$ in the sense of the Fourier convergence theorem, and (iii) vanishes at $x = 0, L$ due to the properties of the sine functions. In fact for any $t > 0$ the series converges very fast to an infinitely differentiable function v (later we will see why) which is therefore the solution to our initial/boundary value problem.

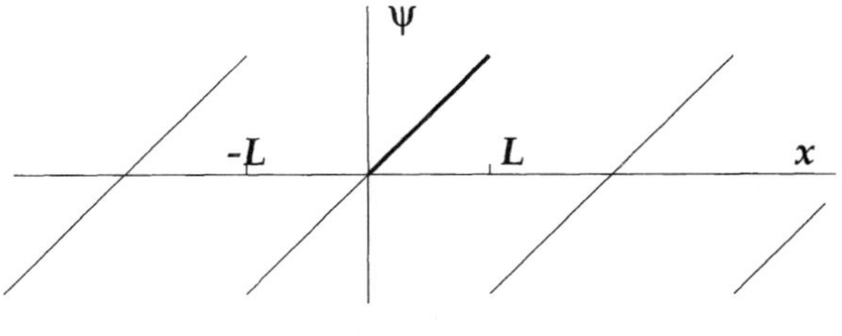

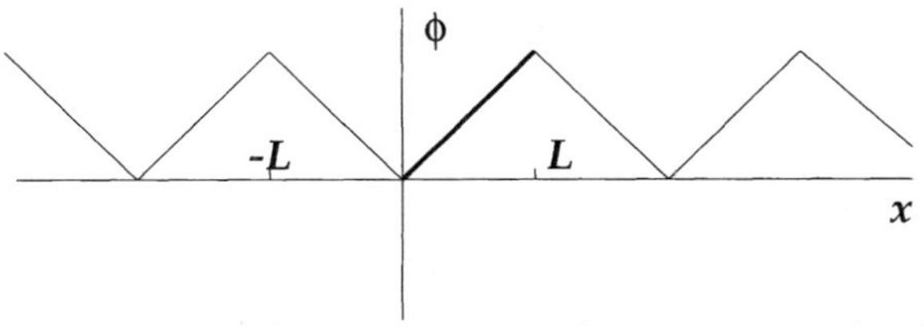

In the no–flow problem we are looking for a solution $u(t, x)$ to the heat equation $u_t = \alpha^2 u_{xx}$ satisfying a given initial condition $u(0, x) = \phi(x),\ 0 < X < L$ and the boundary conditions $u_x(t, 0) = 0 = u_x(t, L),\ t > 0$. In order to find u, we extend ϕ to $(-L, L)$ as an *even* function, then extend it to the entire line by $2L$-periodicity and expand it into the cosine Fourier series. Similarly to the previous case, we find

$$u(t, x) = \frac{a_0}{2} + \sum_{n=1}^{\infty} a_n e^{-\alpha^2 \pi^2 n^2 t / L^2} \cos \frac{\pi n x}{L},$$

where

$$a_n = \frac{2}{L}\int_0^L \phi(x)\cos\frac{\pi n x}{L}\,dx.$$

Due to the property of the cosine factors to have zero derivatives at $x = 0, L$, the function u defined by the series satisfies all the requirements (provided that ϕ is piecewise differentiable).

Examples. (a) *Problem.* A homogeneous bar of length $L = \pi$ and thermal diffusivity $\alpha^2 = 1$ has been maintained at the zero temperature. Starting the moment $t = 0$ the ends of the bar will be maintained at the temperatures 0 and 1. Find the temperature distribution $u(t,x)$ at $t > 0$.

Solution. In view of Example 2.3.1(a) we have $u(t,x) = v(t,x) + x/\pi$ where v satisfies the heat equation $v_t = v_{xx}$, the zero boundary conditions $v(t,0) = 0 = v(t,\pi)$ and the initial condition $v(0,x) = u(0,x) - x/\pi = -x/\pi$. Using the coefficients of the sine Fourier series for the function $\psi(x) = x,\ 0 < x < \pi$, found in Example 2.4.3(a), we compute

$$v(t,x) = \sum_{n=1}^{\infty} \frac{2(-1)^n}{\pi n} e^{-n^2 t}\sin nx,$$

and therefore

$$u(t,x) = \frac{x}{\pi} - \frac{2}{\pi}e^{-t}\sin x + \frac{2}{2\pi}e^{-4t}\sin 2x - \frac{2}{3\pi}e^{-9t}\sin 3x + \ldots$$

In particular, when $t \to \infty$, the temperature distribution approaches the linear function x/π.

(b) *Problem.* A homogeneous bar of length $L = \pi$ and thermal diffusivity $\alpha^2 = 1$ had a linear temperature distribution $\psi(x) = x/\pi$ at the moment $t = 0$ when it was completely insulated. Find the temperature distribution at $t > 0$.

Solution. Using the coefficients of the cosine Fourier series for the function x found in Example 2.4.3(b) we find

$$u(t,x) = \frac{1}{2} - 4\sum_{n\ odd} \frac{e^{-n^2 t}\cos nx}{\pi^2 n^2} =$$

$$\frac{1}{2} - \frac{4}{\pi^2}e^{-t}\cos x - \frac{4}{9\pi^2}e^{-9t}\cos 3x - \frac{4}{25\pi^2}e^{-25t}\cos 5x - \ldots$$

In particular, when $t \to \infty$ the temperature stabilizes at the level $1/2$.

Exercises 2.5.1.

(a) Represent the function $\xi(x) = \sin^2 x$ by cosine and sine Fourier series on the interval $0 < x < \pi$. In the region $0 \le x \le \pi,\ 0 < t < \infty$, write down the solutions to the heat equation $u_t = u_{xx}$ satisfying the initial condition $u(0,x) = \xi(x)$ and (one of) the boundary conditions

$$(g')\ \ u(t,0) = 0 = u(t,\pi), \quad (g'')\ \ u_x(t,0) = 0 = u_x(t,\pi).$$

(b) A homogeneous solid bar of length L and thermal diffusivity α^2 is completely insulated and has the initial temperature distribution x^2 for $0 < x < L$. Find the temperature distribution for $t > 0$.

(c) A homogeneous bar of length L and thermal diffusivity $\alpha^2 = 1$ has the initial temperature distribution

$$\begin{cases} 1 \text{ for } 0 < x < L/2 \\ 0 \text{ for } L/2 < x < L \end{cases}.$$

Find the temperature distribution for $t > 0$ assuming that the ends of the bar are (i') insulated, (i") maintained at zero temperatures.

(d) A homogeneous bar of length L and thermal diffusivity α^2 has zero initial temperature. Extremities of the bar are maintained at the temperature T_0 starting $t = 0$. Find the temperature distribution for $t > 0$.

(e) Compute the infinite sum

$$1 + \frac{1}{16} + \frac{1}{81} + \ldots + \frac{1}{n^4} + \ldots$$

2.5.2. Properties of solutions. The general series solution formulas of the previous section, whether we like them or not, are *the* solutions to our boundary value problems. One can show that solutions to the heat equation with given initial and boundary conditions are unique. It is plausible (although I don't know such a theorem) that the solutions cannot be represented by elementary functions unless the initial condition is a trigonometric polynomial, and in this sense the series cannot be "summed". Instead of complaining about the complexity of the series formula let us try to figure out what it tells us about heat conduction. We will analyze the general solution

$$v(t,x) = \sum_{n=1}^{\infty} b_n e^{-n^2 t} \sin nx, \; b_n = \frac{2}{\pi} \int_0^{\pi} \psi(x) \sin nx \, dx,$$

of the zero–temperature boundary value problem, where we put $L = \pi$, $\alpha^2 = 1$ for simplicity of notation. In the case of the no–flow boundary condition the conclusions would be very similar.

For each particular moment of time $t > 0$ the formula represents the temperature distribution as a superposition of elementary oscillations $\sin nx$ making n half-waves along the bar. The magnitudes of these oscillations are the Fourier coefficients $b_n(t) = b_n e^{-n^2 t}$ of the temperature distribution at the moment t. When t grows, the magnitudes die out exponentially, and — due to the factor n^2 in the exponent — the greater n the faster. The initial condition $\psi(x)$, being piecewise continuous, is bounded in the absolute value by some constant M, and therefore the initial Fourier coefficients $b_n(0) = b_n$ are bounded too: $|b_n(0)| \leq 2M/\pi$. This shows that for any positive t only a few first terms of the series is needed in order to approximate the solution with certain precision, and the greater t the fewer. The conclusion that *contributions of high frequency oscillations* $\sin nx$ *die out much faster than those of low frequency* is one of the most general properties of heat conduction phenomena.

Another manifestation of basically the same property of heat phenomena is their smoothening character: frequent bumps and dips present in the graph of the initial temperature distribution tend to dissolve in the process of heat conduction. This vaguely formulated statement can be transformed to a precise mathematical theorem about solutions of the heat equation: even if the initial condition $v(0,x)$ is discontinuous *the solution* $v(t,x)$ *for* $t > 0$ *is differentiable infinitely many times.*

In order to justify the claim let us differentiate the terms of the series k times in x. We will get a new series (of sine or cosine functions depending on the parity of k) with coefficients $b_n^{(k)}(t) = \pm b_n n^k e^{-n^2 t}$. For each positive t the exponential factor $e^{-n^2 t}$ as a function of n decreases much faster then the power factor n^k increases. In contrast with the Fourier theorem where convergence can be slow, non-absolute and hard to prove, the convergence of the Fourier series with the coefficients $b_n^{(k)}(t)$ for $t > 0$ is absolute and follows immediately from any test (root, ratio, comparison, integral — whichever). It is then easy to show that the term-wise differentiation was legitimate and so the sum of the series is equal to the k-th partial derivative $\partial^k v/\partial x^k$.

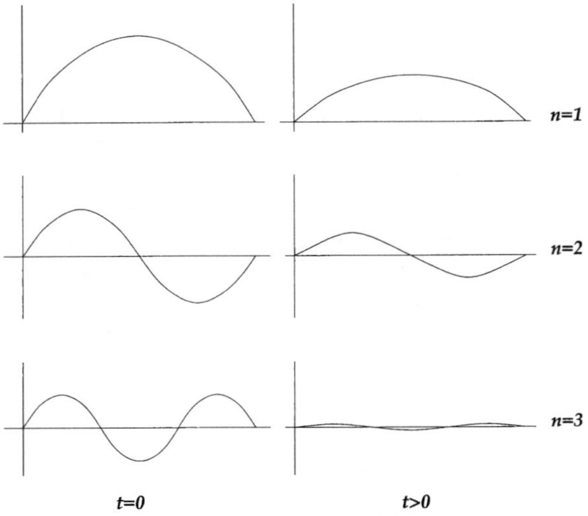

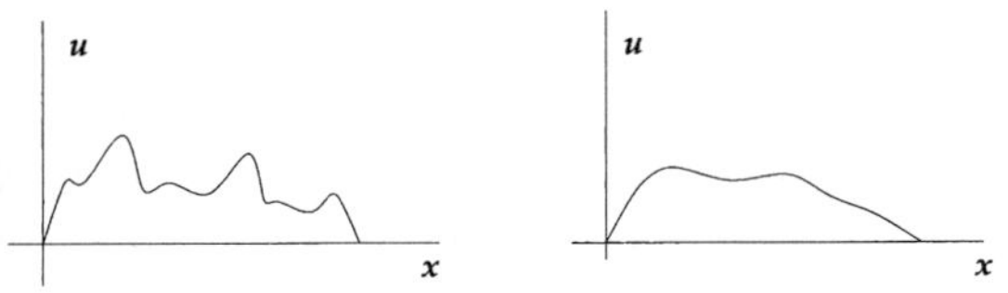

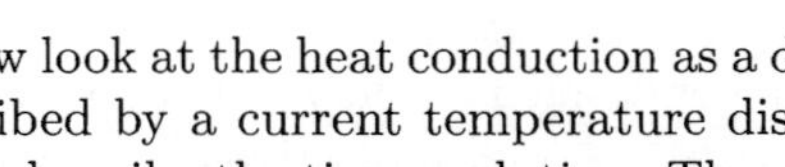

Let us now look at the heat conduction as a dynamical system where the current state is described by a current temperature distribution in the bar, and the heat equation is to describe the time evolution. The solution formula tells us how to find future states of the system via the current state. Does the current state determine the past history of the system?

We have seen examples of initial conditions with Fourier coefficients $b_n(0)$ proportional to $1/n$ or $1/n^2$. For such initial conditions the coefficients $b_n(t) = b_n(0)e^{-n^2 t}$ of the solution series grow with n when $t < 0$. Thus for $t < 0$ the series does not converge at all, and the general solution formula tells us nothing about the temperature distribution in the past. This observation agrees with the following intuitive argument: as a result of heat conduction, different initial temperature distributions eventually become homogeneous, hence the same, or almost the same. Therefore we should be unable to recover the initial distribution from a much later one. Thus heat conduction phenomena are semi-deterministic: the future is determined by the current state while, generally speaking, the past is not.

It turns out that this heuristic principle can be supported by the following precise mathematical statement about the heat equation: *if the initial condition $\psi(x)$ is differentiable only finitely many times, then not only the series solution formula does not make sense for $t < 0$, but no solution to the boundary value problem with the initial condition ψ may exist on any interval $-\varepsilon < t < 0$.*

Indeed, suppose that a function $v(t,x)$ satisfies the heat equation on $-\varepsilon < t \leq 0$ (and is therefore two times differentiable in x for any such t). Considering a moment $t_0 < 0$ as the initial one we can describe the function $v(t,x)$ for $t > t_0$ by the Fourier series formula via the Fourier coefficients for $v(t_0, x)$. Thus the function $v(0,x)$ represented by this series at the moment $t = 0 > t_0$ is differentiable infinitely many times and does not coincide with ψ.

We see that all differentiable initial temperature distributions are going to live forever but many of them do not have any past. This conclusion is a mathematically accurate formulation of the semi-deterministic property of heat conduction.

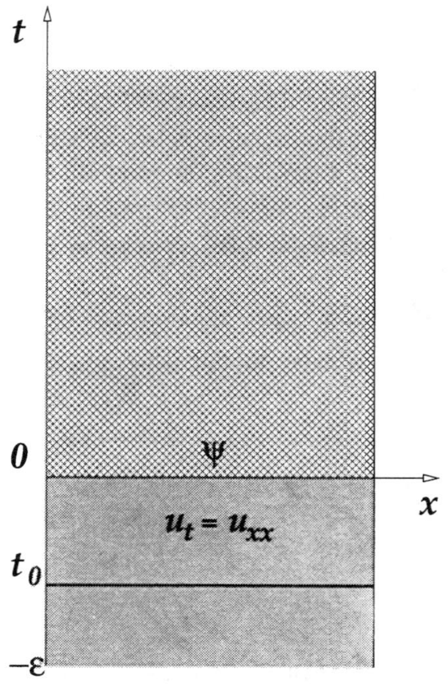

Exercises 2.5.2.

(a) Professor Foulier from the College of Letters and Digits invented new short-term memory hardware. In order to store a string of eight binary digits he suggested to divide a solid bar into 8 equal parts, heat each part to the temperature 1 or cool it to 0 depending on the corresponding digit and then insulate the bar. For instance, by heating the left half of the bar to 1 and cooling the right half to 0, Foulier's device stores the string 11110000 and is still able to read it off after 1 minute. How long will the device be able to store the string 10011001?

(b) Consider the zero-temperature boundary value problem for the heat equation $u_t = u_{xx}$ on the interval $0 \le x \le \pi$. Find the maximal ε such that the solution to the equation exists in the time interval $-\varepsilon < t < 0$ if the initial condition at $t = 0$ is given by the function $\sin^2 x$, $\sin^3 x$.

(c) Returning to diffusion of ink in an infinite tube of water, let us assume that the initial mass distribution of the ink spot is given by a continuous function $m(x)$ vanishing outside some finite interval. Show that the function

$$u(t,x) = \int_{-\infty}^{\infty} \frac{m(\xi)}{\sqrt{4\pi t}} e^{-(x-\xi)^2/4t} d\xi$$

satisfies the diffusion equation $u_t = u_{xx}$ for all x and $0 < t < \infty$. Change the integration variable ξ to $\eta = (x-\xi)/\sqrt{t}$ and show that

$$\lim_{t \to 0+} u(t,x) = m(x)$$

and thus the function u satisfies the initial condition $m(x)$.

(d) Show that the solution $u(t,x)$ in Problem (c) is differentiable infinitely many times at any x and positive t. Deduce that the ink spot cannot evolve backward in time in accordance with the diffusion equation if the initial distribution $m(x)$ is differentiable only finitely many times.

SAMPLE MIDTERM EXAM

1. Expand the 2π-periodic function $f(x)$ equal to 0 for $0 \le x < \pi$ and equal to 1 for $\pi \le x < 2\pi$ into a real Fourier series.

2. Find the solution to the ODE system

$$\dot{x}_1 = 3x_1 - 2x_2$$

$$\dot{x}_2 = 2x_1 - x_2$$

which satisfies the initial condition $x_1(0) = 0,\ x_2(0) = 1$.

3. The temperature distribution in a completely insulated solid bar of length π and of thermal diffusivity $\alpha^2 = 1$ is given by the function $u(0,x) = \cos^2 x - \sin^2 2x$ for $0 \le x \le \pi$. Find the temperature distribution $u(t,x)$ at the moment t.

4. Five linear constant coefficients ODE systems $\dot{\mathbf{x}} = A\mathbf{x}$ have characteristic polynomials $\det(\lambda I - A)$:

$(1):\ \lambda^2 + 2\lambda + 1,\ (2):\ \lambda^2 + \lambda + 1,\ (3):\ \lambda^2 + 1,\ (4):\ \lambda^2 - 2\lambda + 1,\ (5):\ \lambda^2 - 1.$

Four of the systems happened to have the following phase pictures:

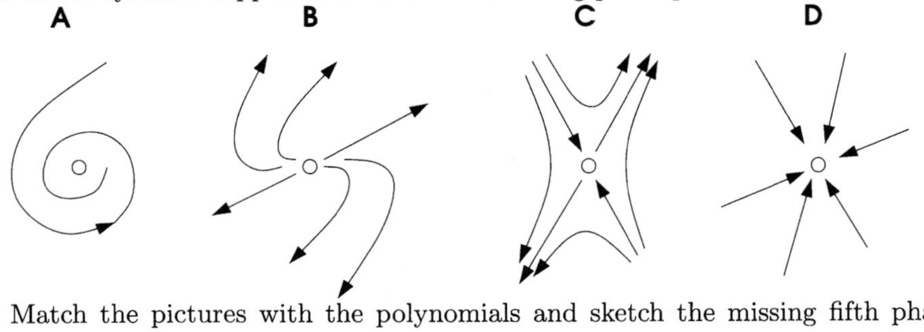

Match the pictures with the polynomials and sketch the missing fifth phase picture.

Which of the five systems have asymptotically stable equilibria at $\mathbf{x} = \mathbf{0}$ and which do not?

CHAPTER 3

Linear Algebra

3.1. Classical problems of Linear Algebra

Nonlinear problems, such as finding maxima or inversion of functions, evaluation of areas and volumes, summation of infinite series, etc. are complicated. Differential and Integral Calculus gives us plenty useful hints how to approach such problems, but simple universal recipes among them are rare. To the contrary, Linear Algebra deals with very simple, linear or quadratic functions. Among numerous questions one may ask about such functions there are, roughly speaking, only four basic, similarly formulated problems which Linear Algebra can handle. It is completeness and simplicity of solutions to these problems what makes Linear Algebra efficient in applications. The four model questions and the answers can be described as follows.

Question 1. *Given m linear functions in n variables,*

$$\begin{array}{lcl} y_1 & = & a_{11}x_1 + ... + a_{1n}x_n \\ & ... & \\ y_m & = & a_{m1}x_1 + ... + a_{mn}x_n \end{array},$$

what is the simplest form to which they can be transformed by linear changes of the variables,

$$\begin{array}{lcl} y_1 & = & b_{11}Y_1 + ... + b_{1m}Y_m \\ ... & & \\ y_m & = & b_{m1}Y_1 + ... + b_{mm}Y_m \end{array}, \quad \begin{array}{lcl} x_1 & = & c_{11}X_1 + ... + c_{1n}X_n \\ ... & & \\ x_n & = & c_{n1}X_1 + ... + c_{nn}X_n \end{array} \quad ?$$

The answer is given by

The Rank Theorem. *Any m linear functions in n variables can be transformed by suitable linear changes of dependent and independent variables to exactly one of the forms:*

$$Y_1 = X_1, ..., Y_r = X_r, Y_{r+1} = 0, ..., Y_m = 0 \text{ where } 0 \le r \le m, n.$$

The number r featuring in the answer is called the **rank** of the set of m linear functions in question.

Question 2. *Given a homogeneous quadratic function in n variables,*

$$Q = q_{11}x_1^2 + 2q_{12}x_1x_2 + 2q_{13}x_1x_3 + ... + q_{nn}x_n^2,$$

what is the simplest form it can be transformed to by a linear change of the variables

$$\begin{array}{lcl} x_1 & = & c_{11}X_1 + ... + c_{1n}X_n \\ & ... & \\ x_n & = & c_{n1}X_1 + ... + c_{nn}X_n \end{array} \quad ?$$

The Inertia Theorem. *Any homogeneous quadratic function in n variables can be transformed by a suitable linear change of the variables to exactly one of the normal forms:*

$$X_1^2 + ... + X_p^2 - X_{p+1}^2 - ... - X_{p+q}^2 \text{ where } 0 \le p + q \le n.$$

The numbers p and q of positive and negative squares in the normal form are called **inertia indices** of the quadratic function in question. If the quadratic function Q is known to be positive everywhere outside the origin, the Inertia Theorem tells us that in a suitable coordinate system Q assumes the form $X_1^2 + ... + X_n^2$ with the inertia indices $p = n$, $q = 0$.

Question 3. Given two homogeneous quadratic functions $Q(x_1, ..., x_n)$ and $S(x_1, ..., x_n)$ of which the first one is known to be positive everywhere outside the origin, what is the simplest form they can be simultaneously transformed to by a linear change of the variables?

The Orthogonal Diagonalization Theorem. *Any pair Q, S of homogeneous quadratic functions in n variables, of which Q is positive everywhere outside the origin, can be transformed by a linear changes of the variables to exactly one of the normal forms*

$$Q = X_1^2 + ... + X_n^2,\ S = \lambda_1 X_1^2 + ... + \lambda_n X_n^2, \text{ where } \lambda_1 \geq ... \geq \lambda_n.$$

Question 4. Given a constant coefficient system of n linear homogeneous 1-st order ordinary differential equations

$$\begin{array}{rcl} \dot{x}_1 & = & a_{11}x_1 + ... + a_{1n}x_n \\ \dots & & \\ \dot{x}_n & = & a_{n1}x_1 + ... + a_{nn}x_n \end{array},$$

what is the simplest form to which it can be transformed by a linear change of the phase variables

$$\begin{array}{rcl} x_1 & = & c_{11}X_1 + ... + c_{1n}X_n \\ \dots & & \\ x_n & = & c_{n1}X_1 + ... + c_{nn}X_n \end{array}?$$

The answer to this question is easier to formulate assuming that the coefficients a_{ij} of the system as well as the coefficients c_{ij} in the change of variables are allowed to be complex numbers.

Example. The system of ODEs

$$\begin{array}{rcl} \dot{x}_1 & = & \lambda x_1 + x_2 \\ \dot{x}_2 & = & \lambda x_2 + x_3 \\ & \dots & \\ \dot{x}_{m-1} & = & \lambda x_{m-1} + x_m \\ \dot{x}_m & = & \lambda x_m \end{array}$$

is equivalent to the single m-th order ODE

$$(\frac{d}{dt} - \lambda)^m y(t) = 0,$$

$$y = x_1,\ \frac{d}{dt}y - \lambda y = x_2,\ (\frac{d}{dt} - \lambda)^2 y = x_3,\ ...,$$

and is called the Jordan cell of size m with the eigenvalue λ. Let us introduce a Jordan system of several Jordan cells of sizes $m_1, ..., m_r$ with eigenvalues $\lambda_1, ..., \lambda_r$ similarly equivalent to the system

$$(\frac{d}{dt} - \lambda_1)^{m_1} y_1 = 0,\ ...,\ (\frac{d}{dt} - \lambda_r)^{m_r} y_r = 0$$

of r unlinked ODEs of orders $m_1, ..., m_r$.

The Jordan Theorem. *Any constant coefficient system of n linear 1-st order ODEs can be transformed by a complex linear changes of phase variables to exactly one (up to reordering of cells) of the Jordan systems with $m_1 + ... + m_r = n$.*

Note that the classification list in the Jordan Theorem (as well as in the Orthogonal Diagonalization Theorem) is not discrete since Jordan systems depend on the choice of complex numbers $\lambda_1, ..., \lambda_r$. In fact the numbers can be found as the roots of the characteristic polynomial $\det(\lambda I - A)$ of the coefficient matrix $A = [a_{ij}]$ of the original ODE system. In the typical case when all roots are simple all Jordan cells have size 1. Thus we arrive at the following corollary of the Jordan Theorem:

A typical constant coefficient system of n linear 1-st order ODEs can be transformed by linear changes of phase variables to the form

$$\dot{X}_1 = \lambda_1 X_1, \ ..., \ \dot{X}_n = \lambda_n X_n.$$

That's about it. One may ask many other similarly looking questions, for instance — about simultaneous classification of triples of quadratic forms or pairs of ODE systems. Such problems are considered unsolvable: Linear Algebra helps to solve only those problems which can be reduced to one of the previous four or to their slightly more general variants. The catch here is not in the word *general* but in the word *reduced*: each of the above theorems has numerous equivalent reformulations and corollaries (we have seen this in the example of the Orthogonal Diagonalization Theorem on the plane), and one needs quite a bit of experience in order to recognize the questions which can be reduced to them and rule out those where Linear Algebra is helpless.

There is however one more basic theorem (or better to say — formula) in Linear Algebra which has no resemblance with the above classifications. It answers the question *which substitutions of the form*

$$\begin{aligned} x_1 &= c_{11}X_1 + ... + c_{1n}X_n \\ &\cdots \\ x_n &= c_{n1}X_1 + ... + c_{nn}X_n \end{aligned}$$

are indeed changes of the variables and therefore allow to express $X_1, ..., X_n$ linearly via $x_1, ..., x_n$. It turns out that there exists a remarkable function det of n^2 variables $c_{11}, ..., c_{nn}$ which vanishes if and only if the square matrix $C = [c_{ij}]$ is not invertible. We begin our study of higher dimensional linear algebra with properties of matrices and determinants.

Exercises 3.1.

(a) Formulate The Rank Theorem in the particular case of two linear functions in two variables. Using the theorem classify linear transformations from the (x_1, x_2)-plane to (y_1, y_2)-plane up to linear changes of coordinates in both planes. Prove The Rank Theorem in the case $m = n = 2$.

(b) Formulate The Inertia Theorem in the particular case $n = 2$ and compare the statement with results of Chapter 1.

(c) Show that $X_1^2 + ... + X_n^2$ is the only one of the normal forms of The Inertia Theorem which is positive everywhere outside the origin.

(d) Prove that the special case $n = 2$ of The Orthogonal Diagonalization Theorem is equivalent to the Orthogonal Diagonalization Theorem of Chapter 1.

(e) Using the binomial formula show that that the Jordan cell of size m with the eigenvalue λ can be written as the m-th order ODE

$$y^{(m)} - \binom{m}{1}\lambda y^{(m-1)} + \binom{m}{2}\lambda^2 y^{(m-2)} + ... + (-1)^{m-1}\binom{m}{m-1}\lambda^{m-1}y' + (-1)^m y = 0.$$

(f) Show that $y(t) = e^{\lambda t}(c_0 + tc_1 + ... + c_{m-1}t^{m-1})$ is the general solution to the ODE $(\frac{d}{dt} - \lambda)^m y = 0$.

(g) Specialize the formulation of the Jordan theorem to the case of $n = 2$ linear ODEs $\dot{\mathbf{x}} = A\mathbf{x}$.

(h) Prove that any complex 2×2-matrix is similar to either a diagonal matrix $\begin{bmatrix} \lambda_1 & 0 \\ 0 & \lambda_2 \end{bmatrix}$ or to the Jordan cell $\begin{bmatrix} \lambda & 1 \\ 0 & \lambda \end{bmatrix}$, where $\lambda_1, \lambda_2, \lambda$ are complex numbers.

3.2. Matrices and Determinants

In our brief introduction to arrays of numbers and their determinants we do not specify what kind of *numbers* we use: real, complex or rational numbers will do equally well.

3.2.1. Matrix Algebra. A rectangular array of numbers with m rows and n columns is called an $m \times n$-matrix. The entry of the matrix A located on the intersection of the i-th row with the j-th column is denoted a_{ij}.

The sum $C = A + B$ of $m \times n$-matrices A and B is an $m \times n$-matrix defined by addition of corresponding entries: $c_{ij} = a_{ij} + b_{ij}$. A scalar multiple $C = \lambda A$ of an $m \times n$-matrix A is similarly defined by $c_{ij} = \lambda a_{ij}$.

The matrix product AB is defined only when the number of columns of A is equal to the number of rows of B.

Example. (a) The product **ab** of the $1 \times n$-matrix $\mathbf{a} = [a_1, ..., a_n]$ and the $n \times 1$-matrix $\mathbf{b} = \begin{bmatrix} b_1 \\ .. \\ b_n \end{bmatrix}$ is defined as the 1×1-matrix $c = a_1 b_1 + ... + a_n b_n$.

More generally, the product $C = AB$ of an $m \times n$-matrix A with an $n \times l$-matrix B is an $m \times l$-matrix C defined in such a way that the entry c_{ij} in the i-th row and j-th column is equal to the product of the i-th row of A with the j-th column of B:

$$c_{ij} = a_{i1} b_{1j} + ... + a_{in} b_{nj} = \sum_{k=1}^{n} a_{ik} b_{kj}, \quad i = 1, ..., m, \; j = 1, ..., l.$$

Examples. (b) The $n \times n$-matrix

$$I_n = \begin{bmatrix} 1 & 0 & ... & 0 \\ 0 & 1 & ... & 0 \\ & & ... & \\ 0 & ... & 0 & 1 \end{bmatrix}$$

is called the identity matrix and satisfies $AI_n = A$, $I_n B = B$. The products $I_n A$ and BI_n are defined only when $m = n$ and $n = l$.

(c) Substitution of n linear forms in l variables,

$$y_1 = b_{11} x_1 + ... + b_{1l} x_l, \; ..., \; y_n = b_{n1} x_1 + ... + b_{nl} x_l,$$

into m linear forms in n variables,

$$z_1 = a_{11} y_1 + ... + a_{1n} y_n, \; ..., \; z_m = a_{m1} y_1 + ... + a_{mn} y_n,$$

yields m linear forms in l variables,

$$z_1 = c_{11} x_1 + ... + c_{1l} x_l, \; ..., \; z_m = c_{m1} x_1 + ... + c_{ml} x_l,$$

with the coefficient matrix C equal to the matrix product AB.

The associative and distributive rules of matrix arithmetics familiar from the theory of matrices of sizes ≤ 2 remain true for arbitrary sizes:

$$(AB)C = A(BC), \; (A + B)C = AC + BC, \; C(A + B) = CA + CB$$

whenever the sizes of A, B, C allow to form the expressions. We leave verification of these formulas to the reader.

A square $n \times n$-matrix A is called **invertible** if there exists an $n \times n$-matrix B such that $AB = I_n = BA$. If such a B exists, it is unique. Indeed, for another matrix B' satisfying $AB' = I = B'A$ we find $B' = B'I = B'(AB) = (B'A)B = IB = B$. This allows us to introduce the notation A^{-1} for the matrix B (when it exists) and call A^{-1} the **inverse matrix**.

Example. (d) Using the matrix notation $\mathbf{x}$ and $\mathbf{x}'$ for the columns of the variables $x_1, ..., x_n$ and $x'_1, ..., x'_n$, we can encode the linear substitution

$$x_1 = a_{11}x'_1 + ... + a_{1n}x'_n, \quad ..., \quad x_n = a_{n1}x'_1 + ... + a_{nn}x'_n$$

by the matrix product formula $\mathbf{x} = A\mathbf{x}'$. We say that the substitution is a **change of variables** if there exists a linear substitution $\mathbf{x}' = B\mathbf{x}$ inverse to it: $A(B\mathbf{x}) = \mathbf{x}$, $B(A\mathbf{x}') = \mathbf{x}'$. Thus $\mathbf{x} = A\mathbf{x}'$ is a change of variables if and only if the matrix A is invertible, in which case the inverse change of variables is given by the formula $\mathbf{x}' = A^{-1}\mathbf{x}$.

Exercises 3.2.1.

(a) Let

$$A = \begin{bmatrix} 1 & 2 & 3 \\ 1 & -1 & 1 \end{bmatrix}, \quad B = \begin{bmatrix} 1 & 2 \\ 3 & 4 \end{bmatrix}, \quad C = \begin{bmatrix} 1 & 2 \\ -2 & 1 \\ 0 & -1 \end{bmatrix}.$$

Compute those of the products

$$(AB)C, A(BC), (BA)C, B(AC), (BC)A, B(CA), (CB)A, C(BA), (CA)B, C(AB), (AC)B, A(CB).$$

which are defined.

(b) Verify the statement of Example (c).

(c) Prove that matrix multiplication is associative: $(AB)C = A(BC)$.

(d) Check the distributive laws of matrix algebra.

(e) Show that if AB is defined then B^tA^t is also defined and is equal to $(AB)^t$. (Here ^{t} is the operation of matrix **transposition** defined by reflecting the arrays about the principal diagonal.)

(f) A square matrix $A = [a_{ij}]$ is called **upper-triangular** if $a_{ij} = 0$ for all $i < j$ and **lower-triangular** if $a_{ij} = 0$ for all $i > j$. Prove that products of upper-triangular matrices are upper-triangular and products of lower-triangular matrices are lower-triangular.

(g) A square matrix $A = [a_{ij}]$ is called **diagonal** if $a_{ij} = 0$ for all $i \neq j$. Which diagonal matrices are invertible?

(h) Let A, B be invertible $n \times n$ matrices. Prove that AB is also invertible and $(AB)^{-1} = B^{-1}A^{-1}$.

(i) If AB is invertible, does it mean that A, B and BA are invertible? Consider separately the cases of square and rectangular matrices A, B.

3.2.2. The determinant function. Let $A = \begin{bmatrix} a_{11} & \ldots & a_{1n} \\ & \ldots & \\ a_{n1} & \ldots & a_{nn} \end{bmatrix}$ be a square $n \times n$ matrix. The determinant $\det A$ is a number defined by the formula

$$\det A = \sum_{\sigma} \varepsilon(\sigma) a_{1\sigma(1)} a_{2\sigma(2)} \ldots a_{n\sigma(n)}.$$

Here σ is a permutation of the indices $1, 2, \ldots, n$. A permutation σ can be considered as an invertible function $i \mapsto \sigma(i)$ from the set of n elements $\{1, \ldots, n\}$ to itself. We use the functional notation $\sigma(i)$ in order to specify the i-th term in the string $(\sigma(1), \ldots, \sigma(n))$ of n indices reordered by σ. Thus each elementary product in the determinant formula contains exactly one matrix element from each row, and these elements are chosen from n different columns. The sum is taken over all $n!$ ways of making such choices. The coefficient $\varepsilon(\sigma)$ in front of the elementary product equals 1 or -1 and is called the sign of the permutation σ. We will explain the general rule of the signs after the following examples.

Examples. (a) For $n = 1$ the determinant of $A = [a_{11}]$ equals a_{11}.
(b) For $n = 2$

$$\begin{vmatrix} a_{11} & a_{12} \\ a_{21} & a_{22} \end{vmatrix} = a_{11}a_{22} - a_{12}a_{21}.$$

(c) For $n = 3$ we have $3! = 6$ summands

$$\begin{vmatrix} a_{11} & a_{12} & a_{13} \\ a_{21} & a_{22} & a_{23} \\ a_{31} & a_{32} & a_{33} \end{vmatrix} =$$

$$a_{11}a_{22}a_{33} - a_{12}a_{21}a_{33} + a_{12}a_{23}a_{31} - a_{13}a_{22}a_{31} + a_{13}a_{21}a_{32} - a_{11}a_{23}a_{32}$$

corresponding to the permutations $(123), (213), (231), (321), (312), (132)$. The rule of signs is schematically shown on the picture.

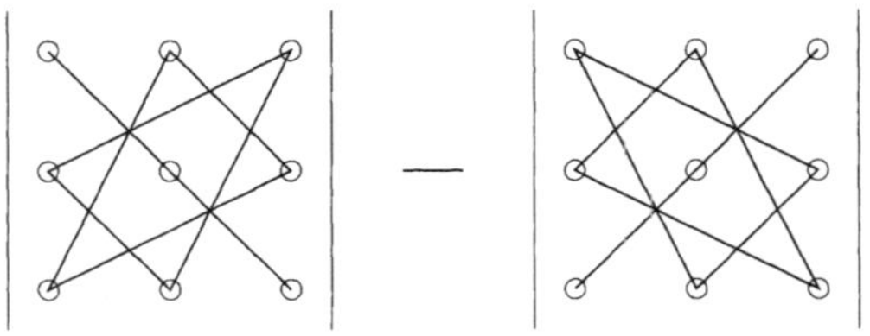

The rule of signs is based on some important properties of permutations. We summarize the properties here and leave their proof to curious readers (see Exercises).

- Each permutation σ on $\{1, \ldots, n\}$ can be represented as a composition $\sigma = \tau_1 \ldots \tau_N$ of transpositions τ which swap two elements and leave the other elements in their places.
- The way of representing σ as a sequence of transpositions is not unique, and different representations of the same σ may have different lengths N but the *parity* of N does not depend on the choice of the representation.
- We put $\varepsilon(\sigma) = (-1)^N$. Permutations represented by even number of transpositions are called even and have $\varepsilon(\sigma) = 1$, and those represented by odd number of transpositions are called odd and have $\varepsilon(\sigma) = -1$. The result of

composing two permutations is even if both are even or both are odd, and it is odd if one is even and the other is odd: $\varepsilon(\sigma\sigma') = \varepsilon(\sigma)\varepsilon(\sigma')$.
- We say that σ **inverses** the pair of indices $i < j$ if $\sigma(i) > \sigma(j)$. The total number $l(\sigma)$ of pairs $i < j$ which σ inverses is called the **length** of the permutation σ. Even (odd) permutations have even (odd) length: $\varepsilon(\sigma) = (-1)^{l(\sigma)}$.

Examples. (d) The permutation (4321) inverses all the 6 pairs of indices and has therefore length $l = 6$. Thus the elementary product $a_{14}a_{23}a_{32}a_{41}$ will occur in the definition of 4×4-determinants with the sign $\varepsilon = (-1)^6 = +1$.

(e) Any transposition is odd (why?) That is why the term $a_{12}a_{21}$ occurs in 2×2-determinants with the negative sign.

(f) Permutations inverse to each other have the same parity since their composition is the identity permutation which is even. This shows that the definition of determinants can be rewritten "by columns":

$$\det A = \sum_{\sigma} \varepsilon(\sigma) a_{\sigma(1)1} \dots a_{\sigma(n)n}.$$

Indeed, each summand in this formula is equal to the summand in the original definition corresponding to the permutation σ^{-1}, and vice versa.

(g) The permutations $(123), (213), (231), (321), (312), (132)$ have lengths $l = 0, 1, 2, 3, 2, 1$ and respectively — signs $\varepsilon = +1, -1, +1, -1, +1, -1$. Notice that each next permutation here is obtained from the previous one by an extra flip.

Exercises 3.2.2.

(a) Compute $\det(\lambda I - A)$ where

$$A = \begin{bmatrix} 0 & 1 & 1 \\ 1 & 0 & -1 \\ -1 & -1 & 0 \end{bmatrix}.$$

Compute determinants of diagonal square matrices.

(b) Do the products $a_{13}a_{24}a_{23}a_{41}a_{35}$, $a_{21}a_{13}a_{34}a_{55}a_{42}$ occur in the definition of determinants of order 5?

(c) Find the signs of the elementary products $a_{23}a_{31}a_{42}a_{56}a_{14}a_{65}$, $a_{32}a_{43}a_{14}a_{51}a_{66}a_{25}$ in the definition of determinants of order 6.

(d) List all the 24 permutations of $\{1, 2, 3, 4\}$, find the length and the sign of each of them.

(e) Prove that the identity permutation is the only permutation of $\{1, 2, ..., n\}$ which has length $l = 0$. What is the maximal length $l(\sigma)$ of permutations on the set $\{1, ..., n\}$?

(f) Show that any transposition τ has odd length $l(\tau)$.

(g) Let σ be a permutation of length $l > 0$. Show that in the string $(\sigma(1), ..., \sigma(n))$ there is a pair $\sigma(i), \sigma(i+1)$ of nearby terms such that $\sigma(i) > \sigma(i+1)$.

(h) Show that composing σ with the transposition

$$\tau^{(i)} = (1, ..., i-1, i+1, i, i+2, ..., n)$$

of nearby indices reduces the length by 1 if $\sigma(i) > \sigma(i+1)$:

$$l(\sigma\tau^{(i)}) = l(\sigma) - 1.$$

Deduce that any permutation σ can be written as a composition of $l(\sigma)$ transpositions $\tau^{(1)}, ..., \tau^{(n-1)}$ of nearby indices.

(i) Show that composing σ with the transposition $\tau^{(i)}$ increases the length by 1 if $\sigma(i) < \sigma(i+1)$:

$$l(\sigma\tau^{(i)}) = l(\sigma) + 1.$$

Deduce that for any two permutations σ, σ' the length $l(\sigma\sigma')$ has the same parity as the sum $l(\sigma) + l(\sigma')$.

(j) Deduce from previous exercises that for compositions $\sigma = t_1 ... t_N$ of any transpositions $l(\sigma)$ and N have the same parity, and that the sign $\varepsilon(\sigma) = \pm 1$ of permutations defined as $(-1)^{l(\sigma)} = (-1)^N$ satisfies

$$\varepsilon(\sigma\sigma') = \varepsilon(\sigma)\varepsilon(\sigma').$$

3.2.3. Properties of determinants. Reflection of a matrix $A = [a_{ij}]$ about the principal diagonal produces the matrix $A^t = [a_{ji}]$ called transposed to A.

(i) *Transposed square matrices have equal determinants:* $\det A^t = \det A$. This follows from Example 3.2.2(f).

We will formulate below properties of determinants in terms of the columns $\mathbf{a}_i$ of the $n \times n$-matrix $A = [\mathbf{a}_1, ..., \mathbf{a}_n]$. The same properties are true for rows as well since transposition of A transforms rows to columns without changing the determinant.

(ii) *Interchanging any two columns changes the sign of the determinant:*

$$\det[..., \mathbf{a}_j, ..., \mathbf{a}_i, ...] = -\det[..., \mathbf{a}_i, ..., \mathbf{a}_j, ...].$$

Indeed, the operation replaces each permutation in the definition of determinants by its composition with the transposition of the indices i and j and thus changes the parity of the permutation.

This property of the determinant considered as a function of n columns is called total antisymmetry. It shows that a matrix with two equal columns has zero determinant. It also allows us to formulate further column properties of determinants referring to the first column only since the properties of all columns are alike.

(iii) *Multiplication of a column by a number multiplies the determinant by this number:*

$$\det[\lambda \mathbf{a}_1, \mathbf{a}_2, ..., \mathbf{a}_n] = \lambda \det[\mathbf{a}_1, \mathbf{a}_2, ..., \mathbf{a}_n].$$

Indeed, this operation simply multiplies each of the $n!$ elementary products by the factor of λ.

This property shows that a matrix with a zero column has zero determinant.

(iv) *The determinant function is* additive *with respect to each column:*

$$\det[\mathbf{a}_1' + \mathbf{a}_1'', \mathbf{a}_2, ..., \mathbf{a}_n] = \det[\mathbf{a}_1', \mathbf{a}_2, ..., \mathbf{a}_n] + \det[\mathbf{a}_1'', \mathbf{a}_2, ..., \mathbf{a}_n].$$

Indeed, each elementary product contains exactly one factor picked from the 1-st column and thus splits into the sum of two elementary products $a'_{\sigma(1)1}a_{\sigma(2)2}...a_{\sigma(n)n}$ and $a''_{\sigma(1)1}a_{\sigma(2)2}...a_{\sigma(n)n}$. Summing up over all permutations yields the sum of two determinants on the right hand side of the formula.

The properties (iv) and (iii) together mean that the determinant function is *linear with respect to each column* separately. Together with the property (ii) they show that *adding a multiple of one column to another one does not change the determinant of the matrix.* Indeed,

$$\det[\mathbf{a}_1 + \lambda \mathbf{a}_2, \mathbf{a}_2, ...] = \det[\mathbf{a}_1, \mathbf{a}_2, ...] + \lambda \det[\mathbf{a}_2, \mathbf{a}_2, ...] = \det[\mathbf{a}_1, \mathbf{a}_2, ...]$$

since the second determinant has two equal columns.

The determinant function shears all the above properties with the identically zero function. The following property shows that these functions do not coincide

(v) $\det I = 1$.

Indeed, since all off-diagonal entries of the identity matrix are zeroes, the only elementary product in the definition of $\det A$ that survives is $a_{11}...a_{nn} = 1$.

The same argument shows that *the determinant of any diagonal matrix equals the product of the diagonal entries.* It is not hard to generalize the argument in order to see that the determinant of any upper- or lower-triangular matrix (that is a square matrix with zero entries everywhere below, or above the principal diagonal) equals the product of the diagonal entries too. One can also deduce this from the following factorization property valid for block-triangular matrices.

Consider an $n \times n$-matrix $\begin{bmatrix} A & B \\ C & D \end{bmatrix}$ subdivided into four matrices A, B, C, D of sizes $m \times m$, $m \times l$, $l \times m$ and $l \times l$ respectively (where of course $m + l = n$). We will call such a matrix block-triangular if $C = \mathbf{0}$. We claim that

$$\det \begin{bmatrix} A & B \\ \mathbf{0} & D \end{bmatrix} = \det A \ \det D.$$

Indeed, consider a permutation σ of $\{1, ..., n\}$ which sends at least one of the indices $\{1, ..., m\}$ to the other part of the set, $\{m+1, ..., m+l\}$. Then σ must send at least one of $\{m+1, ..., m+l\}$ back to $\{1, ..., m\}$. This means that any elementary product in our $n \times n$-determinant which contains a factor from B must contain also a factor from C and hence vanishes if $C = \mathbf{0}$. Thus only the permutations σ which permute $\{1, ..., m\}$ separately from $\{m+1, ..., m+l\}$ contribute to the determinant in question. Elementary products corresponding to such permutations factor into elementary products from $\det A$ and $\det D$ and eventually add up to the product $\det A \det D$.

Of course, the same factorization holds true if $B = \mathbf{0}$ instead of $C = \mathbf{0}$.

We will use the factorization formula in the proof of the following multiplicative property of determinants.

Theorem. *The determinant of the product of two square matrices is equal to the product of their determinants:* $\det(AB) = (\det A)(\det B)$.

Proof. Consider the auxiliary $2n \times 2n$ matrix $\begin{bmatrix} A & 0 \\ -I & B \end{bmatrix}$ with the determinant equal to the product $\det A \det B$ according to the factorization formula. We begin to change the matrix by adding to the last n columns linear combinations of the first n columns with such coefficients that the submatrix B is eventually replaced by zero submatrix. Thus, in order to kill the entry b_{kj} we must add the b_{kj}-multiple of the k-th column to the $n+j$-th column. According to the properties of determinants (see (iv)) these operations do not change the determinant but transform the matrix to the form $\begin{bmatrix} A & C \\ -I & \mathbf{0} \end{bmatrix}$. We leave the reader to check that the entry c_{ij} of the submatrix C in the upper right corner equals $a_{i1}b_{1j} + ... + a_{in}b_{nj}$ so that $C = AB$ is the matrix product! Now, interchanging the i-th and $n+i$-th columns, $i = 1, ..., n$, we change the determinant by the factor of $(-1)^n$ and transform the matrix to the form $\begin{bmatrix} C & A \\ \mathbf{0} & -I \end{bmatrix}$. The factorization formula applies again and yields $\det C \det(-I)$. We conclude that $\det C = \det A \det B$ since $\det(-I) = (-1)^n$ compensates for the previous factor $(-1)^n$.

Corollary. *Invertible matrices have non-zero determinants.*

Indeed, $\det A \det A^{-1} = \det I = 1$ and hence $\det A \neq 0$. The converse statement — that matrices with non-zero determinants are invertible — is also true due to the explicit formula for the inverse matrix to be described in the next section.

Exercises 3.2.3.

(a) Compute the determinants

$$\begin{vmatrix} 13547 & 13647 \\ 28423 & 28523 \end{vmatrix}, \quad \begin{vmatrix} 246 & 427 & 327 \\ 1014 & 543 & 443 \\ -342 & 721 & 621 \end{vmatrix}.$$

(b) Professor Foulier writes his office and home phone numbers as a 7×1-matrix O and a 1×7-matrix H respectively. Help him to compute $\det(OH)$.

(c) How does a determinant change if all the columns are rewritten in the opposite order?

(d) Solve the equation

$$\begin{vmatrix} 1 & x & x^2 & \dots & x^n \\ 1 & a_1 & a_1^2 & \dots & a_1^n \\ 1 & a_2 & a_2^2 & \dots & a_2^n \\ & & & \dots & \\ 1 & a_n & a_n^2 & \dots & a_n^n \end{vmatrix} = 0,$$

where all $a_1, ..., a_n$ are distinct.

(e) A square matrix $A = [a_{ij}]$ is called anti-symmetric if $A^t = -A$ (that is if $a_{ij} = -a_{ji}$ for all i, j). Prove that if n is odd then any anti-symmetric $n \times n$-matrix has zero determinant.

(f) Prove that similar matrices have equal characteristic polynomials:

$$\det(\lambda I - A) = \det(\lambda I - C^{-1}AC).$$

(g) Give another, more conceptual proof of the Theorem. Namely, show first that any function $f[\mathbf{a}_1, ..., \mathbf{a}_n]$ of n columns linear in each of them has the form $\sum a_{i_1 1} ... a_{i_n n} f[\mathbf{e}_{i_1}, ..., \mathbf{e}_{i_n}]$. Deduce that such a function, if totally antisymmetric, must be proportional to the determinant function: $f[\mathbf{a}_1, ..., \mathbf{a}_n] = \det[\mathbf{a}_1, ..., \mathbf{a}_n] f[\mathbf{e}_1, ..., \mathbf{e}_n]$. Next, show that the function $\det[B\mathbf{a}_1, ..., B\mathbf{a}_n]$, where B is a fixed square matrix, is linear in each $\mathbf{a}_i$ and is totally antisymmetric. Deduce that $\det[B\mathbf{a}_1, ..., B\mathbf{a}_n] = \det[\mathbf{a}_1, ..., \mathbf{a}_n] \det[B\mathbf{e}_1, ..., B\mathbf{e}_n]$. Finally, let $\mathbf{a}_i$ be the columns of A. Show that $B\mathbf{a}_i$ are the columns of BA, $B\mathbf{e}_i$ are the columns of B and conclude that $\det(BA) = (\det A)(\det B)$.

3.2.4. Cofactors. In the determinant formula for an $n \times n$-matrix A each elementary product $\pm a_{1\sigma(1)}...$ begins with one of the entries $a_{11}, ..., a_{1n}$ of the first row. The sum of all terms containing a_{11} in the 1-st place is the product of a_{11} with the determinant of the $(n-1) \times (n-1)$-matrix obtained from A by crossing out the 1-st row and the 1-st column. Similarly, the sum of all terms containing a_{12} in the 1-st place looks like the product of a_{12} with the determinant obtained by crossing out the 1-st row and the 2-nd column of A. In fact it differs by the factor of -1 from this product, since switching the columns 1 and 2 changes signs of all terms in the determinant formula and interchanges the roles of a_{11} and a_{12}. Proceeding in this way with $a_{13}, ..., a_{1n}$ we arrive at the cofactor expansion formula for $\det A$ which can be stated as follows.

The $(n-1)$-determinant of the submatrix in A obtained by crossing out the row i and the column j is called the (ij)-minor of A. We denote it M_{ij}. The (ij)-cofactor A_{ij} of the matrix A is the number which differs from the minor M_{ij} by the factor ± 1: $A_{ij} = (-1)^{i+j} M_{ij}$. The chess-board of the signs $(-1)^{i+j}$ is shown on the picture. With these notations, the cofactor expansion formula reads:

$$\det A = a_{11}A_{11} + a_{12}A_{12} + ... + a_{1n}A_{1n}.$$

Example. (a)

$$\begin{vmatrix} a_{11} & a_{12} & a_{13} \\ a_{21} & a_{22} & a_{23} \\ a_{31} & a_{32} & a_{33} \end{vmatrix} = a_{11} \begin{vmatrix} a_{22} & a_{23} \\ a_{32} & a_{33} \end{vmatrix} - a_{12} \begin{vmatrix} a_{21} & a_{23} \\ a_{31} & a_{33} \end{vmatrix} + a_{13} \begin{vmatrix} a_{21} & a_{22} \\ a_{31} & a_{32} \end{vmatrix}.$$

Using the properties (i) and (ii) of determinants we can adjust the cofactor expansion to the i-th row or j-th column:

$$\det A = a_{i1}A_{i1} + ... + a_{in}A_{in} = a_{1j}A_{1j} + ... + a_{nj}A_{nj}, \quad i, j = 1, ..., n.$$

These formulas reduce evaluation of $n \times n$-determinants to that of $(n-1) \times (n-1)$-determinants and can be useful in recursive computations.

Furthermore, we claim that applying the cofactor formula to the entries of the i-th row but picking the cofactors of another row we get zero sum:

$$a_{i1}A_{j1} + ... + a_{in}A_{jn} = 0 \text{ if } i \neq j.$$

Indeed, construct a new matrix $\tilde{A}$ replacing the j-th row by a copy of the i-th row. This "forgery" does not change the cofactors $A_{j1}, ..., A_{jn}$ (since the j-th row is crossed out anyway) and yields the cofactor expansion $a_{i1}A_{j1} + ... + a_{in}A_{jn}$ for $\det \tilde{A}$. But $\tilde{A}$ has two identical rows and hence — zero determinant. The same arguments applied to the columns yield the dual statement

$$a_{1i}A_{1j} + ... + a_{ni}A_{nj} = 0 \text{ if } i \neq j.$$

All the above formulas can be summarized in a single matrix identity. Let us introduce the $n \times n$-matrix $adj(A)$ adjoint to A by placing the cofactor A_{ij} on the intersection of j-th row and i-th column (in other words, $adj(A) = [A_{ij}]^t$).

Example. (b) $adj \begin{bmatrix} a & b \\ c & d \end{bmatrix} = \begin{bmatrix} d & -b \\ -c & a \end{bmatrix}.$

Cofactor Theorem. $A\ adj(A) = (\det A)\ I = adj(A)\ A$.

Corollary. *If* $\det A \neq 0$ *then* A *is invertible and* $A^{-1} = (\det A)^{-1}\ adj(A)$.

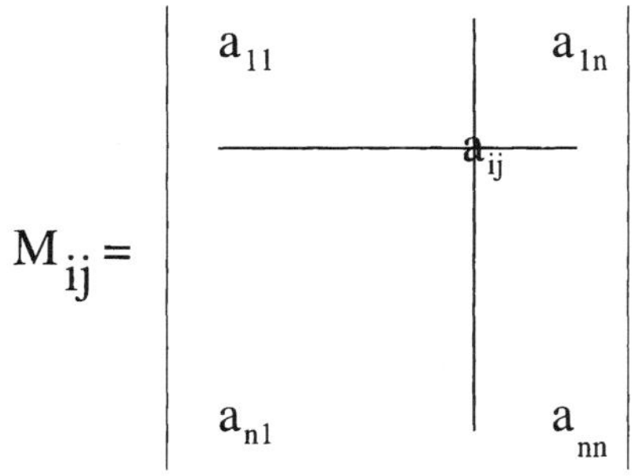

Minors

$$A_{ij} = (-1)^{i+j} M_{ij}$$

i \ j	1	2	3	4	5
1	+	-	+	-	+
2	-	+	-	+	-
3	+	-	+	-	+
4	-	+	-	+	-
5	+	-	+	-	+

$$(-1)^{i+j}$$

Cofactors

Exercises 3.2.4.

(a) Compute determinants using cofactor expansions:

$$\begin{vmatrix} 1 & 2 & 2 & 1 \\ 0 & 1 & 0 & 2 \\ 2 & 0 & 1 & 1 \\ 0 & 2 & 0 & 1 \end{vmatrix},$$

$$\begin{vmatrix} 2 & -1 & 0 & 0 \\ -1 & 2 & -1 & 0 \\ 0 & -1 & 2 & -1 \\ 0 & 0 & -1 & 2 \end{vmatrix}.$$

(b) Compute inverses of the following matrices using the Cofactor Theorem:

$$\begin{bmatrix} 1 & 2 & 3 \\ 3 & 1 & 2 \\ 2 & 3 & 1 \end{bmatrix}, \begin{bmatrix} 1 & 1 & 1 \\ 0 & 1 & 1 \\ 0 & 0 & 1 \end{bmatrix}.$$

(c) Compute

$$\begin{bmatrix} 1 & -1 & 0 & 0 \\ 0 & 1 & -1 & 0 \\ 0 & 0 & 1 & -1 \\ 0 & 0 & 0 & 1 \end{bmatrix}^{-1}.$$

(d) Using Cramer's rule solve the systems of linear equations $A\mathbf{x} = \mathbf{b}$ where A is one of the matrices of Exercise (b) and $\mathbf{b} = [1, 0, 1]^t$.

(e) Express $\det(adj(A))$ of the adjoint matrix via $\det A$.

We conclude our introduction to determinants by an application to systems of linear equations. Let $A\mathbf{x} = \mathbf{b}$ be a system of n linear equations in n unknowns $x_1, ..., x_n$ written in the matrix form, and let $\mathbf{a}_1, ..., \mathbf{a}_n$ be the columns of A.

Corollary. (Cramer's rule.) *If* $\det A \neq 0$, *then the system of linear equations* $A\mathbf{x} = \mathbf{b}$ *has a unique solution given by the formulas:*

$$x_1 = \frac{\det[\mathbf{b}, \mathbf{a}_2, ..., \mathbf{a}_n]}{\det[\mathbf{a}_1, ..., \mathbf{a}_n]}, ..., x_n = \frac{\det[\mathbf{a}_1, ..., \mathbf{a}_{n-1}, \mathbf{b}]}{\det[\mathbf{a}_1, ..., \mathbf{a}_n]}.$$

Indeed, when $\det A \neq 0$, the matrix A is invertible, and the matrix equation $A\mathbf{x} = \mathbf{b}$ implies $\mathbf{x} = A^{-1}\mathbf{b}$. Thus the solution is unique, and $x_i = (\det A)^{-1}(A_{1i}b_1 + ... + A_{ni}b_n)$ according to the cofactor formula for the inverse matrix. But the sum $b_1 A_{1i} + ... + b_n A_{ni}$ is the cofactor expansion for $\det[\mathbf{a}_1, ..., \mathbf{a}_{i-1}, \mathbf{b}, \mathbf{a}_{i+1}, ..., \mathbf{a}_n]$ with respect to the i-th column.

Remark. In the case $n = 2$ Cramer's rule coincides with the determinant formulas for solutions of linear systems mentioned in 1.5.2. The use of these nice-looking formulas for numerical solutions of linear systems with $n > 3$ unknowns is not recommended.

3.3. Vectors and linear systems

3.3.1. 3D and beyond. We do not have to abandon our common sense and everyday experience in order to encounter higher-dimensional spaces.

Examples. (a) The space of visual colors has dimension 3. Indeed, any color can be mixed from red, yellow, blue and is specified by corresponding 3 intensities. *The space of* tastes *has dimension* 4: each taste is specified by intensities of *sweet, salty, sour* and *bitter*.

(b) *Efficient management.* Maximization of profit by means of reallocation of available resources under existing constraints: this task, when expressed mathematically, usually reduces to the problem of finding the maximum of a given linear function on a given polyhedral region. The dimension of the region equals the number of parameters the manager is able to control. When this number does not exceed 3, the maximum is most surely achieved by firing the manager.

(c) Positions of 11 soccer players in the field can be represented by 11 points in the plane. The position of the whole team is therefore a point in the 22-dimensional space. For the same reason, the phase space of n competing species has dimension n. The mass-spring system consisting of n masses connected by springs oscillates in the $6n$-dimensional phase space since the current position in space and the current velocity vector of each mass is needed in order to determine the motion unambiguously.

(d) Physical colors (or sounds) form a space of infinite dimension: the intensity of each contributing frequency should be specified. Similarly, the phase space of the heat equation is infinite-dimensional since the current temperature distribution is characterized by an infinite sequence of Fourier coefficients.

Relying on the visual experience we could easily extend our geometrical approach to vectors in the plane to the case of vectors in the space, but our geometrical intuition seems to fail beyond $3D$. There are two ways out: we can try to train our geometrical intuition to percept higher-dimensional images, or we can substitute for geometrical images their algebraic or analytic expressions (such as coordinates of vectors) and rely only on conclusions derived by coordinate computations. In a sense, we intend to do both. Namely,

- we define n-vectors as columns $\mathbf{x} = \begin{bmatrix} x_1 \\ \dots \\ x_n \end{bmatrix}$ of n numbers;
- introduce componentwise addition and multiplication by scalars:
$$\mathbf{x} + \mathbf{y} = \begin{bmatrix} x_1 + y_1 \\ \dots \\ x_n + y_n \end{bmatrix}, \quad \alpha\mathbf{x} = \begin{bmatrix} \alpha x_1 \\ \dots \\ \alpha x_n \end{bmatrix};$$
- and denote by $\mathbb{R}^n$ the set of all n-vectors provided with these algebraic operations.

Now on all mathematical facts about n-vectors are going to rely on these definitions and should be derived from them by formal algebraic arguments. At the same time, working with algebraically defined objects, we intend to develop a terminological scheme which is deliberately geometrical and appealing to our visual intuition. To begin, we call $\mathbb{R}^n$ the space of n-vectors (or the coordinate n-space, or the Euclidean n-space) and, depending on a context, will often refer to elements of $\mathbb{R}^n$ as *radius-vectors* or *points* in this space. Eventually, when geometrically formulated and algebraically verified facts about the n-space accumulate, one usually discovers that geometrical intuition is as much supportive in higher dimensions as it is in $2D$ or $3D$. Attempting to break the barrier, let us compare geometry of 3- and 4-dimensional cubes.

Examples. (e) Consider the cube in $\mathbb{R}^3$ consisting of all points with coordinates $-1 \leq x_i \leq 1$, $i = 1, 2, 3$. It has 8 vertices $(\pm 1, \pm 1, \pm 1)$ and 6 faces situated in the planes $x_i = \pm 1$, $i = 1, 2, 3$. The centers of the faces are points with coordinates $\pm(1,0,0), \pm(0,1,0), \pm(0,0,1)$. They can be considered as vertices of the **octahedron**, a regular polyhedron inscribed into the cube. The octahedron and the cube have the same symmetries: any linear transformation of the space that preserves the cube preserves therefore the inscribed octahedron, and vice versa. Consider now the 4 vertices of the cube with the product of coordinates equal to 1, that is $(1,1,1), (-1,-1,1), (-1,1,-1), (1,-1,-1)$. We claim that the polyhedron with these vertices is a **regular tetrahedron**. Indeed, all pairwise distances between these 4 points are the same and equal to $2\sqrt{2}$ (for instance, the distance between the 2-nd and 3-rd points is the length of the vector $(0,-2,2)$ which is $\sqrt{0+4+4} = \sqrt{8}$). The angles $\pi/3$ between any two edges in each face of the tetrahedron can be found from inner products of corresponding vectors.

(f) Consider now the cube in $\mathbb{R}^4$ defined by the inequalities $-1 \leq x_i \leq 1$, $i = 1, 2, 3, 4$. It has 16 vertices $(\pm 1, \pm 1, \pm 1, \pm 1)$ and 8 faces of dimension 3 situated in the $3D$-planes $x_i = \pm 1$ and isometric (= "identical") to the $3D$-cube each. Centers of the $3D$-faces are the unit vectors situated on the 4 pairwise perpendicular coordinate axes: $\pm(1,0,0,0), \pm(0,1,0,0), \pm(0,0,1,0), \pm(0,0,0,1)$ (see the last picture on the nect page). They are the vertices of the 4-dimensional "octahedron", a regular polyhedron inscribed into the $4D$-cube which actually has 16 faces isometric to a $3D$-tetrahedron. Let us now examine the vertices with the product of coordinates equal to 1. We should be surprised to find out that there are 8 such vertices, $\pm(1,1,1,1)$, $\pm(1,1,-1,-1)$, $\pm(1,-1,1,-1)$,$\pm(1,-1,-1,1)$, so that they do not form a 4-dimensional analogue of the tetrahedron (the latter would have 5 vertices). The 8 vertices are split into 4 pairs of opposite vectors, have the same distance $\sqrt{1+1+1+1} = 2$ to the origin, and the 4 lines carrying these 4 pairs of vectors are perpendicular to each other. Indeed, all the 6 corresponding inner products are zeroes: $1{\cdot}1+1{\cdot}1+1{\cdot}(-1)+1{\cdot}(-1) = 0$, $1{\cdot}1+1{\cdot}(-1)+(-1){\cdot}1+(-1){\cdot}(-1) = 0$, etc. Therefore the $4D$-polyhedron formed by these 8 vertices is isometric to the $4D$-"octahedron" with the vertices $\pm(2,0,0,0), \pm(0,2,0,0), \pm(0,0,2,0), \pm(0,0,0,2)$ (it differs by the scale factor of 2 from the $4D$-"octahedron" formed by the centers of cube's faces). This result could not be predicted on the basis of our geometrical intuition, that is by analogy with the Example (e). It becomes a singular fact of $4D$ geometry when formulated as such and verified by algebra.

Exercises 3.3.1.

(a) How does the length of the diagonal in the n-dimensional unit cube depend on n? Find the angles the diagonal makes with the edges of the cube.

(b) The n-dimensional **simplex** is defined as a subset in $\mathbb{R}^{n+1}$ given by the equation $x_1 + ... + x_{n+1} = 1$ and by the inequalities $x_1, ..., x_{n+1} \geq 0$. Sketch the simplex for $n = 1, 2, 3$. How many edges are there in the n-dimensional simplex? Find angles between the edges.

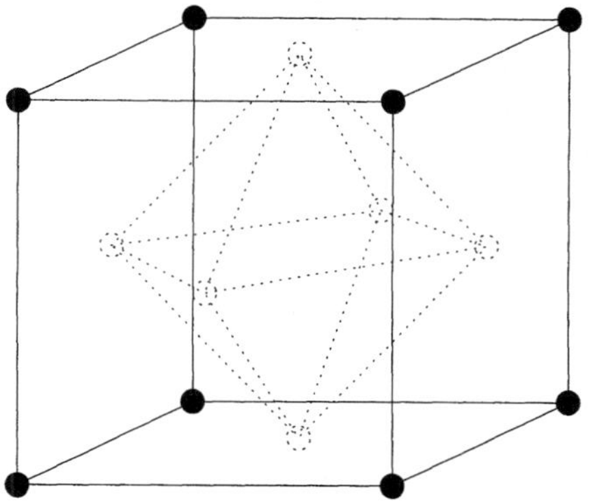

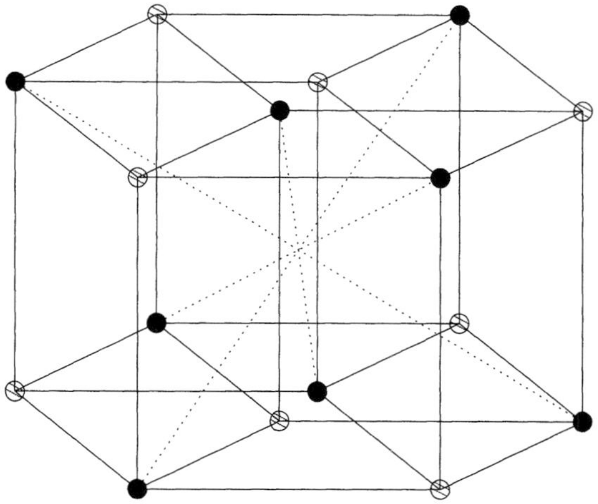

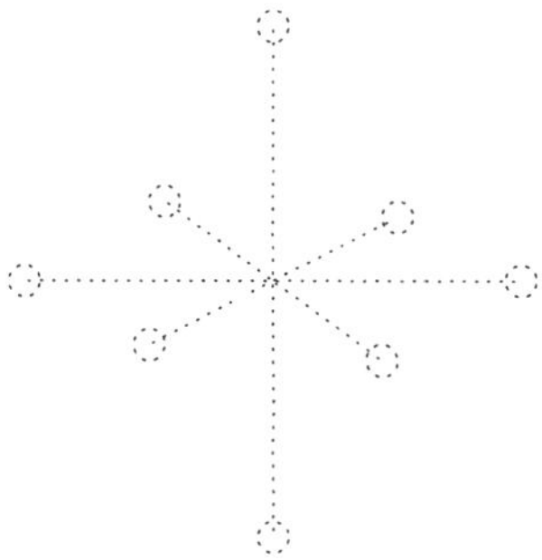

3.3.2. Linear (in)dependence and bases. Any n-vector $\mathbf{x}$ can be uniquely written as a linear combination of the unit coordinate vectors $\mathbf{e}_1, ..., \mathbf{e}_n$:

$$\begin{bmatrix} x_1 \\ ... \\ x_n \end{bmatrix} = x_1 \begin{bmatrix} 1 \\ 0 \\ ... \end{bmatrix} + ... + x_n \begin{bmatrix} ... \\ 0 \\ 1 \end{bmatrix}.$$

A set of n vectors $\mathbf{f}_1, ..., \mathbf{f}_n$ in $\mathbb{R}^n$ is called a basis if any n-vector can be uniquely represented as a linear combination of $\mathbf{f}_1, ..., \mathbf{f}_n$. The conditions *can be* and *uniquely* in this definition deserve separate study.

Let $\mathbf{v}_1, \mathbf{v}_2, ...$ be a set of vectors in $\mathbb{R}^n$. We call this set linearly independent if no vector from $\mathbb{R}^n$ has two different representations as linear combinations of vectors from the set. Another way to say this: no two linear combinations of vectors from the set are equal to each other. Yet another way: if two linear combinations are equal, $\alpha_1 \mathbf{v}_1 + ... + \alpha_N \mathbf{v}_N = \beta_1 \mathbf{v}_1 + ... + \beta_N \mathbf{v}_N$, then their coefficients must be the same: $\alpha_1 = \beta_1, ..., \alpha_N = \beta_N$. Subtracting one of the linear combinations from the other we arrive at a few more reformulations: if $\gamma_1 \mathbf{v}_1 + ... + \gamma_N \mathbf{v}_N = \mathbf{0}$ then necessarily $\gamma_1 = ... = \gamma_N = 0$. In other words, $\mathbf{v}_1, ..., \mathbf{v}_N$ are linearly independent if the vector $\mathbf{0}$ can be written as their linear combination only in the trivial way: $\mathbf{0} = 0\mathbf{v}_1 + ... + 0\mathbf{v}_N$. Equivalently, any non-trivial linear combination of the vectors is not equal to zero: $\gamma_1 \mathbf{v}_1 + ... + \gamma_N \mathbf{v}_N \neq \mathbf{0}$ if at least one of the coefficients $\gamma_i \neq 0$.

Of course, vectors $\mathbf{v}_1, \mathbf{v}_2, ...$ are called linearly dependent if they are not linearly independent. Yet it is useful to have an affirmative reformulation of this condition: the vectors $\mathbf{v}_1, \mathbf{v}_2 ...$ are linearly dependent if and only if some non-trivial linear combination of these vectors equals $\mathbf{0}$: $\gamma_1 \mathbf{v}_1 + ... + \gamma_N \mathbf{v}_N = \mathbf{0}$. The linear combination being non-trivial means that at least one of the coefficients is non-zero. Dividing by this coefficient and moving all other summands to the other side of the equality we obtain the following equivalent formulation: a set of vectors is linearly dependent if and only if one of the vectors is a linear combination of the others. We advise the reader to check at this point that any set containing the vector $\mathbf{0}$ is linearly dependent, any set containing two proportional vectors is linearly dependent, adding new vectors to a linearly dependent set yields a linearly dependent set, any 3 vectors in $\mathbb{R}^2$ are linearly dependent, and 3 vectors in $\mathbb{R}^3$ are linearly dependent only if all three are contained in some plane passing through the origin.

The following statement is a key to vector geometry.

Main Lemma. *Any $n+1$ vectors in $\mathbb{R}^n$ are linearly dependent.*

Proof. Any two vectors on the line $\mathbb{R}^1$ are proportional and therefore linearly dependent. We intend to prove the lemma by deducing from this that any 3 vectors in $\mathbb{R}^2$ are linearly dependent, then deducing from this that any 4 vectors in $\mathbb{R}^3$ are linearly dependent, and so on. Thus we only need to prove that *if* any n vectors in $\mathbb{R}^{n-1}$ are linearly dependent *then* any $n+1$ vectors in $\mathbb{R}^n$ are linearly dependent too. To this end, consider n-vectors $\mathbf{v}_1, ..., \mathbf{v}_{n+1}$ as n-columns. If the last entry in each column is 0, then $\mathbf{v}_1, ..., \mathbf{v}_{n+1}$ are effectively $n-1$-columns, hence some nontrivial linear combination of $\mathbf{v}_1, ..., \mathbf{v}_n$ equals $\mathbf{0}$ and thus the set is linearly dependent. Now consider the case when at least one column has non-zero last entry. Reordering the vectors we may assume that it is the column $\mathbf{v}_{n+1}$. Subtracting the column $\mathbf{v}_{n+1}$ with suitable coefficients $\alpha_1, ..., \alpha_n$ from $\mathbf{v}_1, ..., \mathbf{v}_n$ we form n new columns $\mathbf{u}_1 = \mathbf{v}_1 - \alpha_1 \mathbf{v}_{n+1}, ..., \mathbf{u}_n = \mathbf{v}_n - \alpha_n \mathbf{v}_{n+1}$ which all have the last entries equal to zero. Thus $\mathbf{u}_1, ..., \mathbf{u}_n$ are effectively $n-1$-vectors and are therefore linearly dependent: $\beta_1\mathbf{u}_1 + ... + \beta_n\mathbf{u}_n = \mathbf{0}$ for some $\beta_1, ..., \beta_n$ not all equal to 0. Thus $\beta_1\mathbf{v}_1 + ... + \beta_n\mathbf{v}_n - (\alpha_1\beta_1 + ... + \alpha_n\beta_n)\mathbf{v}_{n+1} = \mathbf{0}$ and hence $\mathbf{v}_1, ..., \mathbf{v}_{n+1}$ are linearly dependent too.

Corollary 1. *Any $m > n$ vectors in $\mathbb{R}^n$ are linearly dependent.*

Let us now analyze the *can be* condition in the definition of a basis. Given a set of vectors $\mathbf{v}_1, \mathbf{v}_2, ...,$ we will denote by $Span(\mathbf{v}_1, \mathbf{v}_2 ...)$ the set of all linear combinations of the vectors from the set. In other words, $Span(\mathbf{v}_1, \mathbf{v}_2, ...)$ consists of all n-vectors which can be represented as linear combinations of $\mathbf{v}_1, \mathbf{v}_2, ...$ For instance, two vectors $\mathbf{v}_1, \mathbf{v}_2$ in $\mathbb{R}^3$ are always contained in some plane P passing through the origin, and $Span(\mathbf{v}_1, \mathbf{v}_2) = P$ if $\mathbf{v}_1$ and $\mathbf{v}_2$ are not proportional to each other. We say that $\mathbf{v}_1, \mathbf{v}_2, ...$ span $\mathbb{R}^n$ (or that $\mathbb{R}^n$ is spanned by $\mathbf{v}_1, \mathbf{v}_2, ...$) if any n-vector *can be* represented as their linear combination: $Span(\mathbf{v}_1, \mathbf{v}_2, ...) = \mathbb{R}^n$.

Corollary 2. *If $k < n$ then $Span(\mathbf{v}_1, ..., \mathbf{v}_k) \neq \mathbb{R}^n$.*

Proof. Indeed, if $Span(\mathbf{v}_1, ..., \mathbf{v}_k) = \mathbb{R}^n$, we can represent the coordinate unit vectors $\mathbf{e}_j$ as linear combinations $\mathbf{e}_j = a_{1j}\mathbf{v}_1 + ... + a_{kj}\mathbf{v}_k$. The coefficient matrix $[a_{ij}]$ has n columns of size k. If $k < n$, the columns are linearly dependent by Corollary 1 and hence $\alpha_1 a_{i1} + ... + \alpha_n a_{in} = 0$ for all $i = 1, ..., k$ and suitable $\alpha_1, ..., \alpha_n$ not all equal 0. Then the non-trivial linear combination $\alpha_1\mathbf{e}_1 + ... + \alpha_n\mathbf{e}_n = (\sum_j \alpha_j a_{1j})\mathbf{v}_1 + ... + (\sum_j \alpha_j a_{kj})\mathbf{v}_k = 0\mathbf{v}_1 + ... + 0\mathbf{v}_k = \mathbf{0}$ in contradiction with linear independence of $\mathbf{e}_1, ..., \mathbf{e}_n$. Thus $Span(\mathbf{v}_1, ..., \mathbf{v}_k) \neq \mathbb{R}^n$.

Returning now to the definition of a basis we see that *any linearly independent set which spans $\mathbb{R}^n$ must consist of exactly n vectors and forms a basis in $\mathbb{R}^n$.*

Bases and coordinate systems in $\mathbb{R}^n$ are in a natural correspondence to each other. Given a basis $\mathbf{f}_1, ..., \mathbf{f}_n$, we can introduce coordinates of a vector $\mathbf{x}$ with respect to this basis as the coefficients $x'_1, ..., x'_n$ of that unique linear combination $x'_1\mathbf{f}_1 + ... + x'_n\mathbf{f}_n$ which is equal to $\mathbf{x}$. The columns of the transition matrix A relating the coordinates with respect to the bases $\mathbf{e}$ and $\mathbf{f}$ as $\mathbf{x} = A\mathbf{x}'$ are exactly the vectors $\mathbf{f}_1, ..., \mathbf{f}_n$ (in the original basis). It is easy to see that expressing the vectors $\mathbf{e}_1, ..., \mathbf{e}_n$ by columns of their coordinates with respect to the basis $\mathbf{f}$ we obtain the matrix inverse to A. Thus bases in $\mathbb{R}^n$ are exactly the n-tuples of columns of invertible $n \times n$-matrices.

Exercises 3.3.2.

(a) Show that any subset of a linearly independent set of vectors is linearly independent.

(b) For each of the 16 subsets in the set of the four 4-vectors

$$\mathbf{v}_1 = (1, 1, -1, -1),\ \mathbf{v}_2 = (1, -1, 1, -1),\ \mathbf{v}_3 = (1, -1, -1, 1),\ \mathbf{v}_4 = (1, 5, -1, -5)$$

find out if the subset is linearly dependent, and if *yes* — represent one of the vectors as a linear combination of the others. Is the set $Span(\mathbf{v}_1, \mathbf{v}_2, \mathbf{v}_3, \mathbf{v}_4)$ finite? Describe $Span(\mathbf{v}_1, \mathbf{v}_2, \mathbf{v}_3, \mathbf{v}_4)$ by a linear equation in $\mathbb{R}^4$ of the form $a_1x_1 + a_2x_2 + a_3x_3 + a_4x_4 = 0$.

(c) Are the 5-vectors

$$(1, 1, 1, 1, 1),\ (1, 2, 4, 8, 16),\ (1, 3, 9, 27, 81),\ (1, 4, 16, 64, 256),\ (1, 5, 25, 125, 625)$$

linearly independent? Do they form a basis in $\mathbb{R}^5$? Why?

(d) Show that the polynomials $1, t, t^2, ..., t^n, ...$ form a linearly independent set. The same — about the functions e^{int}, $n = 0, \pm 1, \pm 2, ...$ Are the functions $\sin t$, $\sin(t + \pi/5)$, $\sin(t + \pi/3)$ linearly dependent? The same — about the functions $\sin^2 t$, $\sin^2(t + \pi/5)$, $\sin^2(t + \pi/3)$.

(e) Let us identify the set of all polynomials $a_0t^n + a_1t^{n-1} + ... + a_n$ of degree $\leq n$ with the space $\mathbb{R}^{n+1}$ of $(n+1)$-tuples of coefficients $(a_0, ..., a_n)$. Let $t_0, ..., t_n$ be distinct numbers. Prove that the degree n polynomials

$$L_0 = (t-t_1)(t-t_2)...(t-t_n),\ L_1 = (t-t_0)(t-t_2)(t-t_3)...(t-t_n),$$

$$L_2 = (t-t_0)(t-t_1)(t-t_3)...(t-n),\ ...,\ L_n = (t-t_0)(t-t_1)...(t-t_{n-1})$$

form a basis in $\mathbb{R}^{n+1}$. Represent the polynomial 1 as a linear combination of $L_0, ..., L_n$.

(f) Is there a polynomial $P(t)$ of degree ≤ 10 which at the points $t = 0, 1, 2, ..., 10$ takes on the values $\sin 0$, $\sin 1$, ..., $\sin 10$?

3.3.3. Subspaces and dimension. *Definition.* A set V of vectors in $\mathbb{R}^n$ is called a linear subspace (or simply subspace) if any linear combinations of vectors from V are also in V.

Examples. (a) The set consisting of one vector $\mathbf{0}$ as well as the set $\mathbb{R}^n$ of all vectors are called trivial subspaces.

(b) Non-trivial subspaces in $\mathbb{R}^2$ are lines passing through the origin, and in $\mathbb{R}^3$ — lines or planes passing through the origin.

(c) The set $Span(\mathbf{v}_1, \mathbf{v}_2, ...)$ is a subspace in $\mathbb{R}^n$ since sums and scalar multiples of linear combinations of $\mathbf{v}_1, \mathbf{v}_2, ...$ are their linear combinations too. The set $Span(\mathbf{v}_1, \mathbf{v}_2, ...)$ is called the subspace spanned by the vectors $\mathbf{v}_1, \mathbf{v}_2, ...$.

Let A be an $m \times n$-matrix. Consider the linear operator $\mathbf{y} = A\mathbf{x}$ from $\mathbb{R}^n$ to $\mathbb{R}^m$ defined by the multiplication of n-columns $\mathbf{x}$ by the matrix. It hass the following linearity property $A(\alpha\mathbf{u} + \beta\mathbf{v}) = \alpha A\mathbf{u} + \beta A\mathbf{v}$ (due to the properties of matrix multiplication). It is easy to see that any function from $\mathbb{R}^n$ to $\mathbb{R}^m$ which satisfies the linearity condition is actually defined by the multiplication by an $m \times n$-matrix. We reserve the term linear transformations for linear operators from $\mathbb{R}^n$ to itself and use the term *linear operators* when the spaces $\mathbb{R}^n$ and $\mathbb{R}^m$ are different, and even in the case $m = n$ if the spaces are considered as two different copies of $\mathbb{R}^n$ rather than the same space.

Examples. (d) The range of a linear operator A from $\mathbb{R}^n$ to $\mathbb{R}^m$ is a subspace in $\mathbb{R}^m$. Indeed, if $\mathbf{u}_1 = A\mathbf{v}_1$ and $\mathbf{u}_2 = A\mathbf{v}_2$ are in the range, then their linear combinations $\alpha_1\mathbf{u}_1 + \alpha_2\mathbf{u}_2 = A(\alpha_1\mathbf{v}_1 + \alpha_2\mathbf{v}_2)$ are in the range too. Applying A to the vectors $\mathbf{e}_1, ..., \mathbf{e}_n$ we see that this subspace is spanned by the columns of the matrix A. The range of a linear operator is often called the column space of the corresponding matrix.

(e) The graph of a linear operator $\mathbf{y} = A\mathbf{x}$ consists of all $n+m$-vectors $(\mathbf{x}, \mathbf{y})$ of the form $(\mathbf{x}, A\mathbf{x})$. The graph is a subspace in $\mathbb{R}^{n+m}$: linear combinations $\alpha(\mathbf{u}, A\mathbf{u}) + \beta(\mathbf{v}, A\mathbf{v}) = (\alpha\mathbf{u} + \beta\mathbf{v}, A(\alpha\mathbf{u} + \beta\mathbf{v}))$ are in the graph.

(f) Consider a linear transformation $\mathbf{x}' = T\mathbf{x}$ from $\mathbb{R}^n$ to itself. Pick a number λ and consider the set of all vectors $\mathbf{x}$ satisfying $A\mathbf{x} = \lambda\mathbf{x}$. They form a subspace in $\mathbb{R}^n$ called the eigenspace of A corresponding to the eigenvalue λ. By definition, all *non-zero* vectors in this subspace are eigenvectors with the eigenvalue λ.

(g) Consider a homogeneous system $A\mathbf{x} = \mathbf{0}$ of m linear equations in n unknowns. The solution set of this system is a subspace in $\mathbb{R}^n$. Indeed, if $A\mathbf{u} = \mathbf{0}$ and $A\mathbf{v} = \mathbf{0}$ then $A(\alpha\mathbf{u} + \beta\mathbf{v}) = \mathbf{0}$. It is called the null-space of the matrix A or the kernel of the corresponding linear operator $\mathbf{y} = A\mathbf{x}$.

(h) The solution set to the inhomogeneous liner system $A\mathbf{x} = \mathbf{b}$ is *not* a subspace (unless $\mathbf{b} = \mathbf{0}$) since $\mathbf{x} = \mathbf{0}$ does not satisfy the system. However, if the system is consistent and $\mathbf{x}_0$ is one of solutions, then any other solution $\mathbf{x}$ differs from $\mathbf{x}_0$ by a vector $\mathbf{v} = \mathbf{x} - \mathbf{x}_0$ from the null-space of A: $A(\mathbf{x} - \mathbf{x}_0) = \mathbf{b} - \mathbf{b} = \mathbf{0}$. Vice versa, adding to $\mathbf{x}_0$ any vector $\mathbf{v}$ from the null-space we obtain another solution to the system: $A(\mathbf{x}_0+\mathbf{v}) = \mathbf{b}+\mathbf{0} = \mathbf{b}$. Thus the solution set to an inhomogeneous system is obtained from the null-space as the translation by the vector $\mathbf{x}_0$. To distinguish from linear subspaces, we will call an affine subspace any set P of vectors obtained from a linear subspace V by translation ($P = V + \mathbf{x}_0$). We will often refer to P as an affine subspace parallel to the linear subspace V (planes in $\mathbb{R}^3$ parallel to a given plane passing through the origin are good examples). Thus, *the solution set to a consistent inhomogeneous system* $A\mathbf{x} = \mathbf{b}$ *of* m *linear equations in* n *unknowns is an affine subspace in* $\mathbb{R}^n$ *parallel to the null-space of* A. Also, *the system* $A\mathbf{x} = \mathbf{b}$ *is consistent if and only if the* m*-vector* $\mathbf{b}$ *is in the column space of* A. This pair of statements is the starting point in the abstract theory of systems of linear algebraic equations.

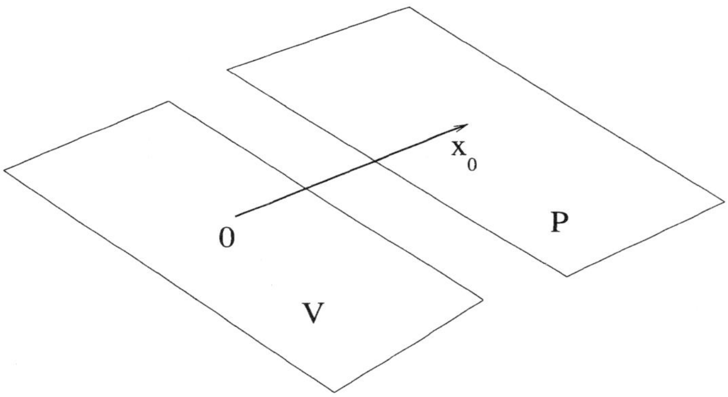

The following proposition shows that "intrinsic" geometry of any non-trivial subspace in $\mathbb{R}^n$ does not differ from geometry of $\mathbb{R}^k$ where k is one of the numbers $1, ..., n-1$ — the dimension of the subspace.

Proposition. *Any non-trivial subspace* V *in* $\mathbb{R}^n$ *is spanned by* $k < n$ *linearly independent vectors.*

Proof. Let $\mathbf{f}_1$ be a non-zero vector in V. If it spans V, we are done (and $k = 1$). If not, there is a vector $\mathbf{f}_2$ in V not proportional to $\mathbf{f}_1$. If $V = Span(\mathbf{f}_1, \mathbf{f}_2)$ then we are done (and $k = 2$). If not, then there is a vector $\mathbf{f}_3$ in V which is not a linear combination of $\mathbf{f}_1$ and $\mathbf{f}_2$. Thus $\mathbf{f}_1, \mathbf{f}_2, \mathbf{f}_3$ are linearly independent, and if they span V, we are done ($k = 3$), and so on. This process cannot continue forever since any n linearly independent vectors in $\mathbb{R}^n$ form a basis and thus span the whole space. Thus, if $V \neq \mathbb{R}^n$, then for some $k < n$ the linearly independent vectors $\mathbf{f}_1, ..., \mathbf{f}_k$ will span V.

Using the vectors $\mathbf{f}_1, ..., \mathbf{f}_k$ as a basis in the subspace V, we write each vector $\mathbf{v}$ from V as a unique linear combination $\mathbf{v} = v_1\mathbf{f}_1 + ... + v_k\mathbf{f}_k$ and thus identify V with the space $\mathbb{R}^k$ of k-columns $\begin{bmatrix} v_1 \\ \dots \\ v_k \end{bmatrix}$. Sums and scalar multiples of vectors from V are expressed by componentwise operations with the k-vectors: if $\mathbf{w} = w_1\mathbf{f}_1 + ... + w_k\mathbf{f}_k$ then $\mathbf{v} + \mathbf{w} = (v_1 + w_1)\mathbf{f}_1 + ... + (v_k + w_k)\mathbf{f}_k$ and $\alpha\mathbf{w} = (\alpha w_1)\mathbf{f}_1 + ... + (\alpha w_k)\mathbf{f}_k$. Thus, the subspace V has exactly the same properties as $\mathbb{R}^k$. In particular, *any linearly independent set of vectors which spans* V *consists of exactly* k *vectors and forms a basis in the subspace.* The number k of vectors in a basis of V is called the dimension of the subspace.

We conclude this section with the following remark which we will need in the proof of the Rank Theorem: any basis $\mathbf{f}_1, ..., \mathbf{f}_k$ in a subspace V can be completed to a basis $\mathbf{f}_1, ..., \mathbf{f}_k, \mathbf{f}_{k+1}, ..., \mathbf{f}_n$ in the whole space $\mathbb{R}^n$ by continuing the procedure described in the proof of the Proposition.

Exercises 3.3.3.

(a) Sketch the subspace in $\mathbb{R}^3$ spanned by the vectors $(1,-1,0),(1,0,-1),(0,1,-1)$. Find the dimension of the subspace in $\mathbb{R}^4$ spanned by the vectors

$$(1,-1,0,1),\ (1,0,-1,1),\ (0,1,-1,0),\ (1,-2,1,1).$$

(b) Find bases in the null-space and range of the linear operator from $\mathbb{R}^4$ to $\mathbb{R}^3$ defined by the matrix

$$\begin{bmatrix} 1 & 0 & 1 & 0 \\ 1 & 1 & 1 & 1 \\ 0 & 1 & 0 & 1 \end{bmatrix}.$$

Complete each of these bases to a basis in $\mathbb{R}^4$ (respectively — $\mathbb{R}^3$).

(c) Find eigenspaces of the linear transformation from $\mathbb{R}^4$ to itself defined by the matrix

$$\begin{bmatrix} 1 & 1 & 0 & 0 \\ 0 & 1 & 1 & 0 \\ 0 & 0 & 1 & 1 \\ 0 & 0 & 0 & 1 \end{bmatrix}.$$

(d) Is a linear subspace affine?

(e) Find the dimension of the affine subspace in $\mathbb{R}^4$ given by the equations $x_1 + x_2 + x_3 = 1$, $x_2 + x_3 + x_4 = 1$, $x_1 - x_4 = 0$. Find the equations of a linear subspace parallel to this affine subspace. Find a basis in this linear subspace and complete this basis to a basis in $\mathbb{R}^4$.

(f) What is the dimension of the graph of a linear operator from $\mathbb{R}^n$ to $\mathbb{R}^m$?

(g) Find a basis in the space of all symmetric traceless 3×3-matrices and in the space of all anti-symmetric 4×4-matrices.

(h) Show that d/dt is a linear operator from the space $\mathbb{R}^{n+1}$ of polynomials $a_n t^n + ... + a_0$ of degree $\leq n$ to the space $\mathbb{R}^{n-1}$ of polynomials of degree $< n$ and find the null-space and the range of this linear operator.

(i) Deduce *from* The Rank Theorem that any subspace in $\mathbb{R}^n$ has a basis, that all such bases have the same number of elements, and that any such a basis can be completed to a basis in $\mathbb{R}^n$.

3.3.4. The Rank Theorem and applications. The matrix of a linear operator $\mathbf{y} = A\mathbf{x}$ from $\mathbb{R}^n$ to $\mathbb{R}^m$ depends on the choice of coordinate systems in $\mathbb{R}^n$ and $\mathbb{R}^m$. The changes $\mathbf{x} = B\mathbf{x}'$, $\mathbf{y} = C\mathbf{y}'$ of coordinated defined by invertible $n \times n$ and $m \times m$ transition matrices B and C yield $\mathbf{y}' = C^{-1}AB\mathbf{x}'$ and thus change the matrix A to $C^{-1}AB$. We will show now that those properties of a linear operator which are independent on the choice of coordinate systems are completely characterized by the dimension r of the range (= the column space of the matrix A) called the rank of the linear operator (and of the matrix).

Example. (a) Consider the $m \times n$-matrix $E_r = \begin{bmatrix} I_r & \mathbf{0} \\ \mathbf{0} & \mathbf{0} \end{bmatrix}$ where $r = 0, 1, ..., \min(m,n)$ (E_r contains the identity matrix of size r in the left upper corner and has all other entries equal to 0). It defines the linear operator given by m linear functions $y_1 = x_1, ..., y_r = x_r, y_{r+1} = 0, ..., y_m = 0$ in n-variables $x_1, ..., x_r, ..., x_n$. The range of this linear operator is the subspace in $\mathbb{R}^m$ given by the equations $y_{r+1} = ... = y_m = 0$ and is spanned by the linearly independent m-vectors $\mathbf{e}_1, ..., \mathbf{e}_r$. Thus the rank of E_r equals r. The null-space of E_r is the subspace in $\mathbb{R}^n$ given by the equations $x_1 = ... = x_r = 0$, and has dimension $n - r$.

Theorem. *A rank r linear operator from $\mathbb{R}^n$ to $\mathbb{R}^m$ has the matrix E_r in suitable coordinate systems in $\mathbb{R}^n$ and $\mathbb{R}^m$.*

Example. (b) The rank of a linear operator from $\mathbb{R}^2$ to another $\mathbb{R}^2$ can be equal to 0, 1 or 2. If it is 0, the linear operator is $\mathbf{0}$. If it is 2, the linear operator identifies the two copies of $\mathbb{R}^2$. In the corresponding coordinate systems, such a linear operator will be described by the identity matrix. If the rank equals 1, the linear operator projects one of the planes onto a line in the other plane.

(c) More generally, the linear operator defined by the matrix E_r can be described as the projection of $\mathbb{R}^n$ along the null-space onto the column space in $\mathbb{R}^m$. The Rank Theorem says that any linear operator A from $\mathbb{R}^n$ to $\mathbb{R}^m$ *is* such a projection along the null-space of A onto the range of A.

Proof of the Rank Theorem. Let $\mathbf{u}_1, ..., \mathbf{u}_r$ be a basis in the range of the linear operator A in question. Complete the basis to a basis of $\mathbb{R}^m$ by choosing suitable vectors $\mathbf{u}_{r+1}, ..., \mathbf{u}_m$ (see the remark in the end of the previous section). Let $\mathbf{v}_1, ..., \mathbf{v}_r$ be any n-vectors such that $\mathbf{u}_1 = A\mathbf{v}_1, ..., \mathbf{u}_r = A\mathbf{v}_r$. Pick a basis $\mathbf{v}_{r+1}, ...$ in the null-space of A. We claim that $\mathbf{v}_1, ..., \mathbf{v}_r, \mathbf{v}_{r+1}, ...$ form a basis in $\mathbb{R}^n$ and thus the linear operator A (which sends $\mathbf{v}_i$ to $\mathbf{u}_i$ for $i \leq r$ and to $\mathbf{0}$ for $i > r$) has in these bases the matrix E_r. In order to justify the claim we will show that any n-vector $\mathbf{x}$ is uniquely written as a linear combination of $\mathbf{v}_i$'s. Indeed, we have $A\mathbf{x} = \alpha_1\mathbf{u}_1 + ... + \alpha_r\mathbf{u}_r$ since $A\mathbf{x}$ is in the range of A. Then $A(\mathbf{x} - \alpha_1\mathbf{v}_1 - ... - \alpha_r\mathbf{v}_r) = \mathbf{0}$ and hence $\mathbf{x} - \alpha_1\mathbf{v}_1 - ... - \alpha_r\mathbf{v}_r = \alpha_{r+1}\mathbf{v}_{r+1} + ...$ since it is in the null-space of A. Thus $\mathbf{x} = \alpha_1\mathbf{v}_1 + ... + \alpha_r\mathbf{v}_r + \alpha_{r+1}\mathbf{v}_r + ...$ On the other hand, if in this equality $\mathbf{x} = \mathbf{0}$, then $A\mathbf{x} = \alpha_1\mathbf{u}_1 + ... + \alpha_r\mathbf{u}_r = \mathbf{0}$ and hence $\alpha_1 = ... = \alpha_r = 0$. Finally the equality $\mathbf{0} = \alpha_{r+1}\mathbf{v}_{r+1} + ...$ implies that $\alpha_{r+1} = ... = 0$ and thus shows that the n-vectors $\mathbf{v}_1, ..., \mathbf{v}_r, \mathbf{v}_{r+1}, ...$ are linearly independent (and in particular — that the total number of these vectors is n).

Corollary. *Any $m \times n$ matrix A of rank r can be transformed to E_r by the transformation $A \mapsto C^{-1}AB$ with suitable invertible matrices B and C.*

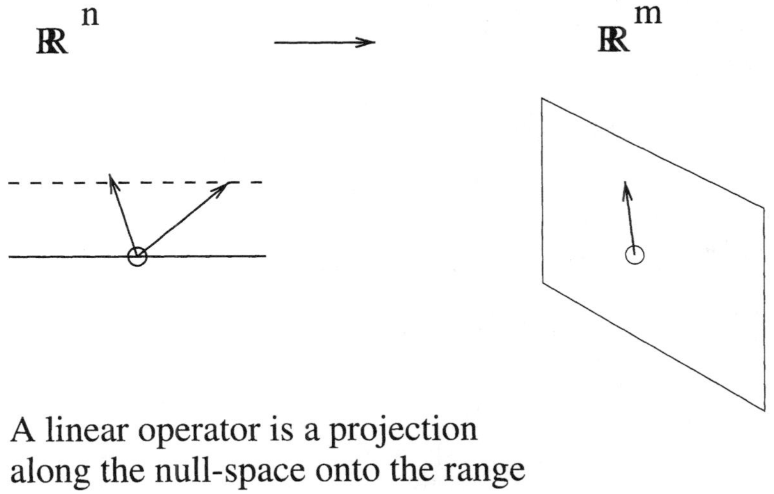

As one can see from applications below The Rank Theorem is a concise summary of a number of simple basic properties of vectors, linear functions, linear operators, subspaces and systems of linear algebraic equations. Some of such properties are already known to us (see Exercises) since they were needed in order to prove the theorem, and some others which may seem new to us can be easily deduced from the theorem.

I. *Linear equations and hyperplanes.*

Let $y(\mathbf{x}) = a_1x_1 + ... + a_nx_n$ be a non-zero linear function. Then the equation $y(\mathbf{x}) = 0$ determines a linear subspace in $\mathbb{R}^n$ of dimension $n-1$. Indeed, y is a linear operator from $\mathbb{R}^n$ to $\mathbb{R}^1$ of rank 1 and has therefore the null-space of dimension $n-1$.

Vice versa, let H be a hyperplane in $\mathbb{R}^n$, that is a linear subspace of dimension $n-1$. *Any hyperplane can be described by one linear equation.* Indeed, let $\mathbf{f}_1, ..., \mathbf{f}_{n-1}$ be a basis in H. Then H is the range of the linear operator from $\mathbb{R}^{n-1}$ to $\mathbb{R}^n$ defined by the matrix whose columns are $\mathbf{f}_1, ..., \mathbf{f}_{n-1}$. According to The Rank Theorem, the linear operator is given by the formulas $Y_1 = X_1, ..., Y_{n-1} = X_{n-1}, Y_n = 0$ in suitable coordinate systems. In particular its range is described by one linear equation $Y_n = 0$.

II. *Several hyperplanes.*

Consider now m linear functions $y_i = a_{i1}x_1 + ... + a_{in}x_n$, $i = 1, ..., m$. We may expect that the intersection of m hyperplanes in $\mathbb{R}^n$ defined by the equations $y_1(\mathbf{x}) = 0, ..., y_m(\mathbf{x}) = 0$ is a subspace of dimension $n-m$. However, if one of the functions, say $y_m(\mathbf{x})$ is a linear combination of the others, then the subspace is actually given by $m-1$ equations $y_1(\mathbf{x}) = ... = y_{m-1}(\mathbf{x}) = 0$ and has dimension greater than $n-m$.

In fact the subspace is the null-space of the linear operator from $\mathbb{R}^n$ to $\mathbb{R}^m$ defined by the functions $y_1(\mathbf{x}), ..., y_m(\mathbf{x})$. According to The Rank Theorem it has dimension $n-r$ where r is the rank of the coefficient matrix $A = [a_{ij}]$.

When the functions $y_1(x), ..., y_m(x)$ are linearly dependent, the range of the linear operator A is contained in some hyperplane $\gamma_1 y_1 + ... + \gamma_m y_m = 0$ and has dimension $r < m$. Vice versa,
if the rank of the linear operator A is smaller than m then the functions $y_1(\mathbf{x}), ...,$ $y_m(\mathbf{x})$ are linearly dependent.

Indeed, according to the Rank Theorem the linear operator is given by the matrix E_r in some new coordinate systems $\mathbf{y}'$ and $\mathbf{x}'$. Each y' is a non-trivial linear combination of y's. When $r < m$, the last row of E_r is zero. This means that the last coordinate function $y'_m(\mathbf{x}')$ — which is equal to some nontrivial linear combination $\gamma_1 y_1 + ... + \gamma_m y_m$ of $y_1(\mathbf{x}), ..., y_m(\mathbf{x})$ — is the identically zero function of $x_1, ..., x_n$.

In particular, *any m linear functions in $n < m$ variables are linearly dependent.*

III. *The row space.*

Linear combinations of the linear functions $y_1(\mathbf{x}), ..., y_m(\mathbf{x})$ form a subspace in the space $\mathbb{R}^n$ of all n-rows. The subspace is often called the row space of the coefficient matrix $A = [a_{ij}]$. The Rank Theorem shows that *the dimension of the row space coincides with the dimension of the column space and is equal to the rank of the matrix* since this is true for the matrices E_r.

The row space of A coincides with the column space of A^t. Thus we can rephrase the above conclusion this way: *transposed matrices have the same rank.*

IV. *Codimension.*

Any subspace V in $\mathbb{R}^m$ of dimension d is the range of some linear operator A from $\mathbb{R}^d$ to $\mathbb{R}^m$. Namely, V is the column space of the $m \times d$-matrix A whose columns $\mathbf{f}_1, ..., \mathbf{f}_d$ form a basis in V. Here is the dual statement: *any subspace V in $\mathbb{R}^m$ of dimension d is given by $c = m - d$ linearly independent linear homogeneous equations.* Indeed, according to The Rank Theorem the linear operator A has the matrix E_d in some new coordinate systems, and hence the range V of the linear operator is given by the equations $y'_{d+1} = ... = y'_m = 0$ in these new coordinates. The number $c = m - d$ is called the **codimension** of the subspace V in $\mathbb{R}^m$. In particular, subspaces of codimension 1 are hyperplanes.

One of the corollaries of the rank theorem is that *for any matrix, the codimension of the null-space coincides with the dimension of the column space and is equal to the rank.* Several applications of this statement are described in the following examples.

V. *Dimension counting.*

Two distinct planes in $\mathbb{R}^3$ passing through the origin intersect along a line passing through the origin, while a line through the origin meets a plane not containing it only at the origin. How do these obvious statements generalize to higher dimensions?

Consider the subspace W in $\mathbb{R}^m$ obtained as the intersection of subspaces U and V of codimensions k and l, that is given respectively by k and l independent linear equations. *If U and V together span the whole ambient space $\mathbb{R}^m$ then their intersection W has codimension $k + l$.* (In particular, if $k + l > m$, the subspaces cannot span the whole m-space.) Equivalently, if $p = m - k$ and $q = m - l$ are dimensions of the subspaces U and V which together span the whole ambient space $\mathbb{R}^m$, then their intersection W has dimension $d = m - k - l = p + q - m$. Indeed, consider linear operators $\mathbf{y} = A\mathbf{u}$ and $\mathbf{y} = B\mathbf{v}$ from $\mathbb{R}^p$ and $\mathbb{R}^q$ to $\mathbb{R}^m$ with the ranges U and V. We define a single linear operator C from $\mathbb{R}^{p+q}$ to $\mathbb{R}^m$: $C(\mathbf{u}, \mathbf{v}) = A\mathbf{u} - B\mathbf{v}$. The range of C is spanned by U and V and thus coincides with $\mathbb{R}^m$. On the other hand, the intersection W consists of all m-vectors $\mathbf{w}$ representable as $\mathbf{w} = A\mathbf{u} = B\mathbf{v}$ and therefore has the same dimension d as the null-space of C. Since codimension of the null-space is equal to the rank, we have $p + q - d = m$.

More generally,
if the subspaces U and V of dimensions p and q span together a subspace in $\mathbb{R}^n$ of dimension m then their intersection W has dimension $d = p + q - m \geq p + q - n$. Indeed, the subspace spanned by U and V can be identified with $\mathbb{R}^m$.

An illustration to the rule ($n = 4$, $k = l = 2$): two distinct 2-dimensional subspaces in $\mathbb{R}^4$ intersect along a line if and only if they both are contained in the same 3-dimensional subspace, and meet only at the origin otherwise.

VI. *The abstract theory of linear algebraic systems.*

Consider a system $A\mathbf{x} = \mathbf{b}$ of m linear equations in n unknowns and denote r the rank of the coefficient matrix A. Then

(i) the solution set of the corresponding homogeneous system $A\mathbf{x} = \mathbf{0}$ is a subspace in $\mathbb{R}^n$ of codimension r (the null-space of A);

(ii) the system $A\mathbf{x} = \mathbf{b}$ is consistent if and only if $\mathbf{b}$ belongs to the subspace in $\mathbb{R}^m$ of dimension r (the column space of A);

(iii) if it does, the solution set is an affine subspace in $\mathbb{R}^n$ of codimension r parallel to the null-space.

Consider now systems $A\mathbf{x} = \mathbf{b}$ with the number of equations m equal to the number of unknowns n (so that A is a square matrix).

(iv) If $\det A \neq 0$, *the system has a unique solution for any* $\mathbf{b}$. *In particular, the homogeneous system* $A\mathbf{x} = \mathbf{0}$ *has only the trivial solution* $\mathbf{x} = \mathbf{0}$.

(v) If $\det A = 0$, *the homogeneous system has non-trivial solutions (which form a linear subspace of dimension* $n - r$*), and the systems* $A\mathbf{x} = \mathbf{b}$ *have no solutions for some* $\mathbf{b}$ *and infinitely many solutions for some others.*

The statements (i) – (v) constitute the final point in the abstract theory of linear algebraic equations. They follow easily from the description, provided by the Rank Theorem, of the linear operator $\mathbf{y} = A\mathbf{x}$ as the projection along the null-space onto the range and from the fact that for a square $n \times n$-matrix A with $\det A = 0$ we have $\det E_r = \det C^{-1} \det A \det B = 0$ and hence $r < n$. In the next section we develop a practical algorithm for solving linear algebraic systems with numerical coefficients.

Exercises 3.3.4.

(a) Using the Rank Theorem classify linear operators from $\mathbb{R}^3$ to $\mathbb{R}^3$ up to changes of coordinates in both spaces. Sketch the pictures illustrating the action of the operators defined by corresponding matrices E_r.

(b) Find a coordinate system in which the linear operator given by the formulas

$$\begin{aligned} y_1 &= 2x_1 - x_2 - x_3, \\ y_2 &= -x_1 + 2x_2 - x_3, \\ y_3 &= -x_1 - x_2 + 2x_3, \end{aligned}$$

has the matrix E_2. Are the functions $y_1(\mathbf{x}), y_2(\mathbf{x}), y_3(\mathbf{x})$ linearly dependent? For which $\mathbf{b}$ the system $y_1(\mathbf{x}) = b_1, y_2(\mathbf{x}) = b_2, y_3(\mathbf{x}) = b_3$ is consistent? has a unique solution?

(c) Find the dimension of the intersection V of the hyperplanes $x_1 + x_2 + x_3 = 0$ and $x_2 + x_3 + x_4 = 0$ in $\mathbb{R}^4$. Are there hyperplanes which intersect V at a point? along a line? a plane? a subspace of dimension 3? If *yes* give examples of such hyperplanes.

(d) List all possible dimensions of intersection of two 3-dimensional subspaces U and V in $\mathbb{R}^5$. Give examples illustrating each possibility. What are the dimensions of subspaces spanned by U and V in your examples?

(e) Find a linear operator from $\mathbb{R}^3$ to $\mathbb{R}^4$ whose range coincides with the hyperplane $x_1 + x_2 + x_3 + x_4 = 0$.

(f) Suppose that the system $A\mathbf{x} = \mathbf{b}$ of m linear equations in 1999 unknowns has a unique solution for $\mathbf{b} = (1, 0, ..., 0)$. Does it imply that

– the null-space of A is trivial?
– the rank of A equals 1999?
– $m \geq 1999$?
– A is invertible?
– A^tA is invertible?
– $\det(AA^t) \neq 0$?
– the rows of A are linearly independent?
– the columns of A are linearly independent?

(g) Given a linear operator A from $\mathbb{R}^n$ to $\mathbb{R}^m$, a linear operator B from $\mathbb{R}^m$ to $\mathbb{R}^n$ is called right inverse to A if $AB = I_m$ and left inverse to A if $BA = I_n$. Prove that a left inverse exists if and only if A has rank m and that the right inverse exist if and only if A has rank n. Is a left (right) inverse unique when exists? Consider separately the case $m = n$ of square matrices.

3.4. Gaussian elimination

It will take your PC another millennium to evaluate the determinant of a 20×20 floating-point matrix using the definition of determinants as sums of 20! elementary products, and — a tiny fraction of a second, using the algorithm described in this section.

3.4.1. Row reduction. Usually, solving a system of linear algebraic equations with coefficients given numerically we express, using one of the equations, the 1-st unknown via the other unknowns and eliminate it from the remaining equations, then express the 2-nd unknown from one of the remaining equations, etc., and finally arrive to an equivalent algebraic system which is easy to solve starting from the last equation and working backward. This computational procedure called Gaussian elimination can be conveniently organized as a sequence of operations with rows of the coefficient matrix of the system. Namely, we use three elementary row operations:

- interchange of two rows;
- division of a row by a non-zero number;
- subtraction of a multiple of one row from another one.

Example. (a) Solving the system

$$\begin{array}{rcrcrcr} & & x_2 & + & 2x_3 & = & 3 \\ 2x_1 & + & 4x_2 & & & = & -2 \\ 3x_1 & + & 5x_2 & + & x_3 & = & 0 \end{array}$$

by Gaussian elimination, we pull the 2-nd equation up (since the 1-st equation does not contain x_1), divide it by 2 (in order to express x_1 via x_2) and subtract it 3 times from the 3-rd equation in order to get rid of x_1 in there. Then we use the 1-st equation (which has become the 2-nd one in our pile) in order to eliminate x_2 from the 3-rd equation. The coefficient matrix of the system is subject to the elementary row transformations:

$$\left[\begin{array}{rrr|r} 0 & 1 & 2 & 3 \\ 2 & 4 & 0 & -2 \\ 3 & 5 & 1 & 0 \end{array}\right] \mapsto \left[\begin{array}{rrr|r} 2 & 4 & 0 & -2 \\ 0 & 1 & 2 & 3 \\ 3 & 5 & 1 & 0 \end{array}\right] \mapsto \left[\begin{array}{rrr|r} 1 & 2 & 0 & -1 \\ 0 & 1 & 2 & 3 \\ 3 & 5 & 1 & 0 \end{array}\right]$$

$$\mapsto \left[\begin{array}{rrr|r} 1 & 2 & 0 & -1 \\ 0 & 1 & 2 & 3 \\ 0 & -1 & 1 & 3 \end{array}\right] \mapsto \left[\begin{array}{rrr|r} 1 & 2 & 0 & -1 \\ 0 & 1 & 2 & 3 \\ 0 & 0 & 3 & 6 \end{array}\right] \mapsto \left[\begin{array}{rrr|r} 1 & 2 & 0 & -1 \\ 0 & 1 & 2 & 3 \\ 0 & 0 & 1 & 2 \end{array}\right].$$

The final "triangular" shape of the coefficient matrix is an example of the row-echelon form. If read from bottom to top, it represents the system $x_3 = 2$, $x_2+2x_3 = 3$, $x_1 + 2x_2 = -1$ which is ready to be solved by back substitution: $x_3 = 2$, $x_2 = 3 - 2x_3 = 3 - 4 = -1$, $x_1 = -1 - 2x_2 = -1 + 2 = 1$. The process of back substitution, expressed in the matrix form, consists of a sequence of elementary operations of the 3-rd type:

$$\left[\begin{array}{rrr|r} 1 & 2 & 0 & -1 \\ 0 & 1 & 2 & 3 \\ 0 & 0 & 1 & 2 \end{array}\right] \mapsto \left[\begin{array}{rrr|r} 1 & 2 & 0 & -1 \\ 0 & 1 & 0 & -1 \\ 0 & 0 & 1 & 2 \end{array}\right] \mapsto \left[\begin{array}{rrr|r} 1 & 0 & 0 & 1 \\ 0 & 1 & 0 & -1 \\ 0 & 0 & 1 & 2 \end{array}\right].$$

The last matrix is an example of the reduced row-echelon form and represents the system $x_1 = 1$, $x_2 = -1$, $x_3 = 2$ which is "already solved".

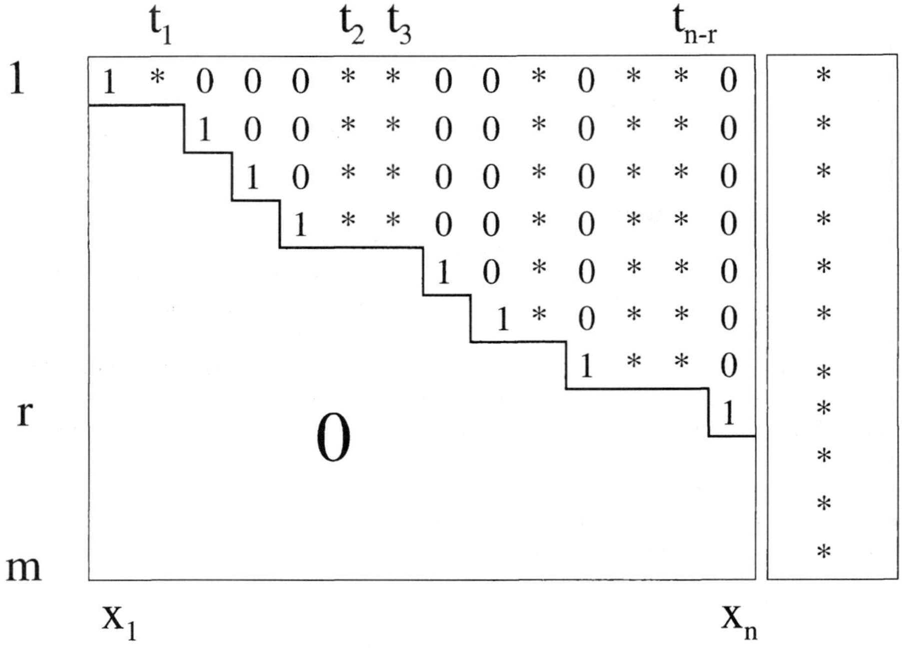

A reduced row-echelon form

In general, Gaussian elimination is an algorithm of reducing an augmented matrix to a row-echelon form by means of elementary row operations. By an augmented matrix we mean simply a matrix subdivided into two blocks $[A|B]$. The augmented matrix of a linear system $A\mathbf{x} = \mathbf{b}$ in n unknowns is $[\mathbf{a}_1, ..., \mathbf{a}_n|\mathbf{b}]$ where $\mathbf{a}_1, ..., \mathbf{a}_n$ are columns of A, but we will also make use of augmented matrices with B consisting of several columns. Operating with rows $[\alpha_1, ..., \alpha_n|\beta_1, ...]$ of augmented matrices we will refer to the leftmost non-zero number among α_i as the leading entry of the row. We say that the augmented matrix $[A|B]$ is in the row-echelon form of rank r if the $m \times n$-matrix A satisfies the following conditions:

- each of the first r rows has the leading entry equal to 1;
- the leading entries of the rows $1, 2, ..., r$ are situated respectively in the columns with indices $j_1, ..., j_r$ satisfying $j_1 < j_2 < ... < j_r$;
- all the rows of A with indices $i > r$ are zero.

Notice that a row-echelon matrix has zero entries everywhere below and to the left of each leading entry. A row-echelon matrix is called reduced if all the entries in the columns $j_1, ..., j_r$ above the leading entries are also equal to zero.

Example. (b) If the matrix A of a linear system is in the row-echelon form and indeed has one or several zero rows in the bottom, then the system contains equations of the form $0x_1 + ... + 0x_n = b$. If at least one of such b is non-zero, the system is inconsistent. If all of them are zeroes, the system is consistent and is ready to solve by back substitution. For instance, the following augmented matrix is in the row-echelon form of rank 2:

$$\left[\begin{array}{ccc|c} 1 & 2 & 3 & 0 \\ 0 & 0 & 1 & 2 \\ 0 & 0 & 0 & 0 \end{array}\right].$$

It corresponds to the system $x_1 + 2x_2 + 3x_3 = 0,\ x_3 = 2,\ 0 = 0$. The system is consistent: $x_3 = 2, x_2 = t, x_1 = -3x_3 - 2x_2 = -6 - 2t$ satisfy the system *for any value of the parameter t*. We see that the presence of leading entries in the columns $j_1, ..., j_r$ of the row-echelon form allows to express the unknowns $x_{j_1}, ..., x_{j_r}$ via the unknowns x_j with $j \neq j_1, ..., j_r$ while the values $t_1, ..., t_{n-r}$ of the unknowns $x_j,\ j \neq j_1, ..., j_r$ are completely ambiguous. The general solution to the linear system depends in this case on the $n - r$ parameters $t_1, ..., t_{n-r}$.

The algorithm of row reduction of an augmented matrix $[A|B]$ to the row-echelon form can be described by the following instructions. Let n be the number of columns of A. Then the algorithm consists of n steps. At the step $l = 1, ..., n$ we assume that the matrix formed by the columns of A with the indices $j = 1, ..., l - 1$ is already in the row-echelon form of some rank $s < l$ with the leading entries located in some columns $j_1 < ... < j_s < l$. The l-th step begins with locating the first non-zero entry in the column l below the row s. If none — the l-th step is over, since the columns $1, ..., l$ are already in the row echelon form of rank s. Otherwise — the first non-zero entry is located in a row $i(> s)$, and we perform the following operations:

(i) interchange the rows i and $s + 1$ of the augmented matrix,

(ii) divide the whole row $s + 1$ of the augmented matrix by the leading entry (it is now $a_{s+1,l} (\neq 0)$),

(iii) annihilate all the entries in the column l below the leading entry of the $s + 1$-st row by subtracting suitable multiples of the $s + 1$-st row of the augmented matrix from all rows with indices $i > s + 1$.

The l-th step is over since the columns $1, ..., l$ are now in the row-echelon form of rank $s + 1$.

When an augmented matrix $[A|B]$ has been reduced to a row-echelon form with the leading entries $a_{1,j_1} = ... = a_{r,j_r} = 1$, the back substitution algorithm, which reduces it further to a reduces row-echelon form, consists of r steps which we number by $l = r, r - 1, ..., 1$ (and perform in this order). On the l-th step we subtract from each of the rows $i = 1, ..., l - 1$ of the augmented matrix the l-th row multiplied by a_{i,j_l} (and thus annihilate all the entries of the column j_l above the leading one).

Exercises 3.4.1.
Solve the systems of linear equations

$$(a)\ \begin{array}{l} 2x_1 - x_2 - x_3 = 4 \\ 3x_1 + 4x_2 - 2x_3 = 11 \\ 3x_1 - 2x_2 + 4x_3 = 11 \end{array},\quad (b)\ \begin{array}{l} x_1 - 2x_2 + x_3 + x_4 = 1 \\ x_1 - 2x_2 + x_3 - x_4 = -1 \\ x_1 - 2x_2 + x_3 + 5x_4 = 5 \end{array},\quad (c)\ \begin{array}{l} x_1 + x_2 - 3x_3 = -1 \\ 2x_1 + x_2 - 2x_3 = 1 \\ x_1 + x_2 + x_3 = 3 \\ x_1 + 2x_2 - 3x_3 = 1 \end{array}$$

$$(d)\ \begin{array}{l} 2x_1 + x_2 + x_3 = 2 \\ x_1 + 3x_2 + x_3 = 5 \\ x_1 + x_2 + 5x_3 = -7 \\ 2x_1 + 3x_2 - 3x_3 = 14 \end{array},\quad (e)\ \begin{array}{l} x_1 - 2x_2 + 3x_3 - 4x_4 = 4 \\ x_2 - x_3 + x_4 = -3 \\ x_1 + 3x_2 - 3x_4 = 1 \\ -7x_2 + 3x_3 + x_4 = -3 \end{array},\quad (f)\ \begin{array}{l} 2x_1 + 3x_2 - x_3 + 5x_4 = 0 \\ 3x_1 - x_2 + 2x_3 - 7x_4 = 0 \\ 4x_1 + x_2 - 3x_3 + 6x_4 = 0 \\ x_1 - 2x_2 + 4x_3 - 7x_4 = 0 \end{array}$$

(g) $\begin{aligned} 3x_1 + 4x_2 - 5x_3 + 7x_4 &= 0 \\ 2x_1 - 3x_2 + 3x_3 - 2x_4 &= 0 \\ 4x_1 + 11x_2 - 13x_3 + 16x_4 &= 0 \\ 7x_1 - 2x_2 + x_3 + 3x_4 &= 0 \end{aligned}$, (h) $\begin{aligned} x_1 + x_2 + x_3 + x_4 + x_5 &= 7 \\ 3x_1 + 2x_2 + x_3 + x_4 - 3x_5 &= -2 \\ x_2 + 2x_3 + 2x_4 + 6x_6 &= 23 \\ 5x_1 + 4x_2 + 3x_3 + 3x_4 - x_5 &= 12 \end{aligned}$

(i) Find those λ for which the system is consistent:

$$\begin{aligned} 2x_1 - x_2 + x_3 + x_4 &= 1 \\ x_1 + 2x_2 - x_3 + 4x_4 &= 2 \\ x_1 + 7x_2 - 4x_3 + 11x_4 &= \lambda \end{aligned}$$

3.4.2. Applications. The row-reduction algorithms allow to find a basis in the null-space, in the row space and in the column space of a given matrix, and to compute efficiently determinants and inverses of square matrices given numerically.

I. *Suppose that an $m \times n$-matrix A has been reduced by elementary row operations to the row-echelon form A' with the leading entries $a_{1,j_1} = \ldots = a_{r,j_r} = 1$, $j_1 < \ldots < j_r$. Then (i) $\operatorname{rk} A = \operatorname{rk} A' = r$, (ii) the first r rows of A' form a basis in the row space of A, (iii) the columns of A with the indices $j_1, \ldots, j_r$ form a basis in the column space of A.*

Indeed, the row operations do not change the row space of the matrix. The non-zero rows of a row-echelon matrix are linearly independent and thus form a basis in the row-space. Row operations do change the columns $[\mathbf{a}_1, \ldots, \mathbf{a}_n]$ of a matrix, but they preserve linear dependencies among them: $\alpha_1\mathbf{a}_1 + \ldots + \alpha_n\mathbf{a}_n = \mathbf{0}$ if and only if $\alpha_1\mathbf{a}_1' + \ldots + \alpha_n\mathbf{a}_n' = \mathbf{0}$. The columns $\mathbf{a}_{j_1}', \ldots, \mathbf{a}_{j_r}'$ containing leading entries of the row echelon matrix A' form a basis in the column space of A', and hence the columns $\mathbf{a}_{j_1}, \ldots, \mathbf{a}_{j_r}$ form a basis in the column space of A.

Example. (a) The following row reduction

$$\begin{bmatrix} 1 & 2 & 3 & -1 \\ 2 & 4 & 5 & 1 \\ 3 & 6 & 8 & 0 \end{bmatrix} \mapsto \begin{bmatrix} 1 & 2 & 3 & -1 \\ 0 & 0 & -1 & 3 \\ 0 & 0 & -1 & 3 \end{bmatrix} \mapsto \begin{bmatrix} 1 & 2 & 3 & -1 \\ 0 & 0 & 1 & -3 \\ 0 & 0 & 0 & 0 \end{bmatrix}$$

shows that the matrix has rank 2, the row space has a basis $(1, 2, 3, -1)$, $(0, 0, 1, -3)$, and the column space has a basis $(1, 2, 3)^t$, $(3, 5, 8)^t$.

II. Suppose that the augmented matrix $[A|\mathbf{b}]$ of the system $A\mathbf{x} = \mathbf{b}$ has been transformed to a reduced row-echelon form $[A'|\mathbf{b}']$ with the leading entries positioned in the columns $j_1 < j_2 < \ldots < j_r$. These columns are the unit coordinate vectors $\mathbf{e}_1, \ldots, \mathbf{e}_r$, and the system is consistent only if $\mathbf{b}'$ is their linear combination, $\mathbf{b}' = b_1'\mathbf{e}_1 + \ldots + b_r'\mathbf{e}_r$. Assuming that it is the case we can assign arbitrary values $t_1, \ldots, t_{n-r}$ to the unknowns x_j, $j \neq j_1, \ldots, j_r$, and express $x_{j_1}, \ldots, x_{j_r}$ as linear inhomogeneous functions of $t_1, \ldots, t_{n-r}$. The general solution to the system will have the form $\mathbf{x} = \mathbf{c}_0 + t_1\mathbf{c}_1 + \ldots + t_{n-r}\mathbf{c}_{n-r}$ of a linear combination of some n-vectors $\mathbf{c}_0, \mathbf{c}_1, \ldots, \mathbf{c}_{n-r}$. We claim that *the vectors $\mathbf{c}_1, \ldots, \mathbf{c}_{n-r}$ form a basis in the null-space of the matrix A.* Indeed, substituting $t = 0$ we see that $\mathbf{c}_0$ is a particular solution to the system and hence $\mathbf{x} - \mathbf{c}_0 = t_1\mathbf{c}_1 + \ldots + t_{n-r}\mathbf{c}_{n-r}$ is the general solution to the homogeneous system $A\mathbf{x} = \mathbf{0}$. In addition, we see that *the general solution to the inhomogeneous system is the affine subspace in $\mathbb{R}^n$ obtained from the null-space by the translation through the vector $\mathbf{c}_0$.*

Example. (b) Consider the system $A\mathbf{x} = \mathbf{0}$ with the matrix A from the example (a). Transform the matrix to the reduced row-echelon form:

$$\ldots \mapsto \begin{bmatrix} 1 & 2 & 3 & -1 \\ 0 & 0 & 1 & -3 \\ 0 & 0 & 0 & 0 \end{bmatrix} \mapsto \begin{bmatrix} 1 & 2 & 0 & 8 \\ 0 & 0 & 1 & -3 \\ 0 & 0 & 0 & 0 \end{bmatrix}.$$

The general solution to the system assumes the form

$$\begin{bmatrix} x_1 \\ x_2 \\ x_3 \\ x_4 \end{bmatrix} = \begin{bmatrix} -2t_1 - 8t_2 \\ t_1 \\ 3t_2 \\ t_2 \end{bmatrix} = t_1 \begin{bmatrix} -2 \\ 1 \\ 0 \\ 0 \end{bmatrix} + t_2 \begin{bmatrix} -8 \\ 0 \\ 3 \\ 1 \end{bmatrix}.$$

The columns $(-2,1,0,0)^t$ and $(-8,0,3,1)^t$ form therefore a basis in the null-space of the matrix A.

III. *Suppose that during the row reduction of an $n \times n$ matrix A to the row-echelon form A' we performed k times the operation of row interchange and applied the operation of division by leading entries $\alpha_1, ..., \alpha_r$. If the rank r of the row-echelon form is smaller than n then* $\det A = 0$. *If $r = n$ then* $\det A = (-1)^k \alpha_1 ... \alpha_n$.

Indeed, each row interchange reverts the sign of the determinant, divisions of a row by α divides the determinant by α, and subtraction of a multiple of one row from another one does not change the determinant. Thus $\det A = (-1)^k \alpha_1 ... \alpha_r \det A'$. The row-echelon matrix is upper-triangular and has n leading 1s on the diagonal when $r = n$. In this case $\det A' = 1$. When $r < n$ we have $\det A' = 0$.

IV. *Given an $n \times n$-matrix A, introduce the augmented matrix $[A|I_n]$ (where I_n is the identity matrix) and transform it to the reduced row-echelon form $[A'|B]$ by elementary row operations. If $A' = I_n$ then $B = A^{-1}$.*

Indeed, the equality $A' = I_n$ means that $\operatorname{rk} A = n$ and thus A^{-1} exists. Then the system $A\mathbf{x} = \mathbf{b}$ has a unique solution for any $\mathbf{b}$, and for $\mathbf{b} = \mathbf{e}_1, ..., \mathbf{e}_n$ the corresponding solutions $\mathbf{x} = A^{-1}\mathbf{e}_1, ..., A^{-1}\mathbf{e}_n$ are the columns of the inverse matrix A^{-1}. These solutions can be found by simultaneous row reduction of the augmented matrices $[A|\mathbf{e}_1], ..., [A|\mathbf{e}_n]$ and thus coincide with the columns of the matrix B in the reduced row-echelon form $[I_n|B]$.

Example. (c) Let us compute $\det A$ and A^{-1} for the matrix of the Example 4.1(a). We have:

$$\left[\begin{array}{ccc|ccc} 0 & 1 & 2 & 1 & 0 & 0 \\ 2 & 4 & 0 & 0 & 1 & 0 \\ 3 & 5 & 1 & 0 & 0 & 1 \end{array}\right] \mapsto \left[\begin{array}{ccc|ccc} 1 & 2 & 0 & 0 & \frac{1}{2} & 0 \\ 0 & 1 & 2 & 1 & 0 & 0 \\ 0 & -1 & 1 & 0 & -\frac{3}{2} & 1 \end{array}\right] \mapsto$$

$$\mapsto \left[\begin{array}{ccc|ccc} 1 & 2 & 0 & 0 & \frac{1}{2} & 0 \\ 0 & 1 & 2 & 1 & 0 & 0 \\ 0 & 0 & 1 & \frac{1}{3} & -\frac{1}{2} & \frac{1}{3} \end{array}\right],$$

where one row interchange and divisions by 2 and by 3 were applied. Thus $\det A = (-1) \cdot 2 \cdot 3 = -6$, and the matrix is invertible. Back substitution eventually yields the inverse matrix:

$$\mapsto \left[\begin{array}{ccc|ccc} 1 & 2 & 0 & 0 & \frac{1}{2} & 0 \\ 0 & 1 & 0 & \frac{1}{3} & 1 & -\frac{2}{3} \\ 0 & 0 & 1 & \frac{1}{3} & -\frac{1}{2} & \frac{1}{3} \end{array}\right] \mapsto \left[\begin{array}{ccc|ccc} 1 & 0 & 0 & -\frac{2}{3} & -\frac{3}{2} & \frac{4}{3} \\ 0 & 1 & 0 & \frac{1}{3} & 1 & -\frac{2}{3} \\ 0 & 0 & 1 & \frac{1}{3} & -\frac{1}{2} & \frac{1}{3} \end{array}\right].$$

Remark. Gaussian elimination algorithms are unlikely to work for matrices depending on parameters. To see why, try the row reduction in order to solve a linear system of the form $(\lambda I - A)\mathbf{x} = \mathbf{0}$ depending on the parameter λ, or — even better — apply Gaussian elimination to the system $a_{11}x_1 + a_{12}x_2 = b_1$, $a_{21}x_1 + a_{22}x_2 = b_2$ depending on the 6 parameters $a_{11}, a_{12}, a_{21}, a_{22}, b_1, b_2$.

Exercises 3.4.2.

For the following matrices, find the rank, a basis in the null-space, a basis in the column space and a basis in the row space:

$$(a)\ \begin{bmatrix} 0 & 4 & 10 & 1 \\ 4 & 8 & 18 & 7 \\ 10 & 18 & 40 & 17 \\ 1 & 7 & 17 & 3 \end{bmatrix} \qquad (b)\ \begin{bmatrix} 14 & 2 & 6 & 8 & 2 \\ 6 & 104 & 21 & 9 & 17 \\ 7 & 6 & 3 & 4 & 1 \\ 35 & 30 & 15 & 20 & 5 \end{bmatrix}$$

$$(c)\ \begin{bmatrix} 1 & 0 & 0 & 1 & 4 \\ 0 & 1 & 0 & 2 & 5 \\ 0 & 0 & 1 & 3 & 6 \\ 1 & 2 & 3 & 14 & 32 \\ 4 & 5 & 6 & 32 & 77 \end{bmatrix} \qquad (d)\ \begin{bmatrix} 2 & 1 & 1 & 1 \\ 1 & 3 & 1 & 1 \\ 1 & 1 & 4 & 1 \\ 1 & 1 & 1 & 5 \\ 1 & 2 & 3 & 4 \\ 1 & 1 & 1 & 1 \end{bmatrix} \qquad (e)\ \begin{bmatrix} 2 & 1 & 3 & -1 \\ 3 & -1 & 2 & 0 \\ 1 & 3 & 4 & -2 \\ 4 & -3 & 1 & 1 \end{bmatrix}$$

For the following matrices, compute the determinant and the matrix inverse:

$$(f)\ \begin{bmatrix} 2 & 2 & -3 \\ 1 & -1 & 0 \\ -1 & 2 & 1 \end{bmatrix} \qquad (g)\ \begin{bmatrix} 1 & 1 & 1 & 1 \\ 1 & 1 & -1 & -1 \\ 1 & -1 & 1 & -1 \\ 1 & -1 & -1 & 1 \end{bmatrix} \qquad (h)\ \begin{bmatrix} 2 & 1 & 0 & 0 \\ 3 & 2 & 0 & 0 \\ 1 & 1 & 3 & 4 \\ 2 & -1 & 2 & 3 \end{bmatrix}$$

(i) Let $P^{(ij)}$, $i \neq j$, denote the $n \times n$-matrix $[p_{kl}]$ with $p_{ij} = p_{ji} = 1$, $p_{ii} = p_{jj} = 0$, all other diagonal entries $p_{kk} = 1$ and all other entries equal to 0. Let $D_i(\lambda)$ be the diagonal $n \times n$-matrix with the i-th diagonal entry equal to λ and all other diagonal entries equal to 1. Let $E_{ij}(a)$, $i \neq j$, be an $n \times n$-matrix with the (ij)-entry equal to a, all diagonal entries equal to 1 and all off-diagonal entries equal to 0. Show that the three elementary row operations on an $n \times n$-matrix A act as the left multiplication of A by suitable matrices P_{ij}, $D_i(\lambda)$ and $E_{ij}(a)$ respectively.

(j) Suppose that the row reduction of a square matrix A to a row-echelon form does not involve the operation of row transposition. Prove that $A = LU$ where L is a lower-triangular square matrix, and U is an upper-triangular square matrix.

3.5. Quadratic forms

Elements of the n-dimensional Euclidean geometry, the theory of quadratic forms and their applications are the subject of this section.

3.5.1. Inertia indices. A quadratic form in $\mathbb{R}^n$ is defined as a homogeneous degree 2 polynomial in n variables:

$$Q(\mathbf{x}) = \sum_{i=1}^{n}\sum_{j=1}^{n} q_{ij}x_i x_j.$$

In this sum, the terms $x_i x_j$ and $x_j x_i$ with $i \neq j$ are similar, and we will assume that the total coefficient $q_{ij} + q_{ji}$ is distributed "equally", $q_{ij} = q_{ji}$, so that the matrix $Q = [q_{ij}]$ of the quadratic form $Q(\mathbf{x})$ is symmetric: $Q^t = Q$. A linear change of coordinates $\mathbf{x} = C\mathbf{x}'$ transforms the quadratic form to a quadratic form $Q'(\mathbf{x}') = \sum_{kl} q'_{kl}x'_k x'_l$ with the coefficient matrix $Q' = C^tQC$. Indeed,

$$Q'(\mathbf{x}') = \sum_{i,j} q_{ij} \sum_{k,l} c_{ik}c_{jl}x'_k x'_l = \sum_{k,l} x'_k x'_l \sum_{i,j} c_{ik}q_{ij}c_{jl}.$$

Notice that the matrix C^tQC is symmetric automatically:
$(C^tQC)^t = C^tQ^tC^{tt} = C^tQC$.

More generally, if $\mathbf{x} = C\mathbf{y}$ is a linear operator from $\mathbb{R}^m$ to $\mathbb{R}^n$, the composite function $Q(C\mathbf{y})$ is a quadratic form in $\mathbb{R}^m$ with the symmetric $m \times m$ matrix C^tQC. Indeed, invertibility of C did not play any role in the previous computation.

Theorem. *Any quadratic form $Q(\mathbf{x})$ in $\mathbb{R}^n$ in a suitable coordinate system $X_1, ..., X_n$ is equal to one of the quadratic forms $\pm X_1^2 \pm ... \pm X_r^2$, $r = 0, ..., n$.*

In order to prove the theorem we associate to a quadratic form $Q(\mathbf{x})$ a new function of two vectors (abusing notation, we denote it by the same letter) :

$$Q(\mathbf{u}, \mathbf{v}) = \sum_{i=1}^{n}\sum_{j=1}^{n} u_i q_{ij} v_j.$$

It is symmetric, $Q(\mathbf{u}, \mathbf{v}) = Q(\mathbf{v}, \mathbf{u})$, since the coefficient matrix is symmetric. It is bilinear,

$$Q(\mathbf{u}, \alpha\mathbf{v} + \beta\mathbf{w}) = \alpha Q(\mathbf{u}, \mathbf{v}) + \beta Q(\mathbf{u}, \mathbf{w}),\ Q(\alpha\mathbf{v} + \beta\mathbf{w}, \mathbf{u}) = \alpha Q(\mathbf{v}, \mathbf{u}) + \beta Q(\mathbf{w}, \mathbf{u}),$$

since $Q(\mathbf{u}, \mathbf{x})$ with $\mathbf{u}$ fixed is a linear function of $x_1, ..., x_n$. The function $Q(\mathbf{u}, \mathbf{v})$ is called the symmetric bilinear form associated with the quadratic form $Q(\mathbf{x})$. The associated quadratic and bilinear forms determine each other by the formulas:

$$Q(\mathbf{x}) = Q(\mathbf{x}, \mathbf{x}),\ Q(\mathbf{u}, \mathbf{v}) = \frac{1}{2}[Q(\mathbf{u} + \mathbf{v}) - Q(\mathbf{u}) - Q(\mathbf{v})].$$

The coefficients q_{ij} of the quadratic form Q in a coordinate system with the basis $\mathbf{f}_1, ..., \mathbf{f}_n$ can be computed as the values $q_{ij} = Q(\mathbf{f}_i, \mathbf{f}_j)$ of the associated bilinear form.

Example. (a) The quadratic forms $\pm x_1^2 \pm ... \pm x_r^2$ are associated with the bilinear forms $(\mathbf{u}, \mathbf{v}) = \pm u_1v_1 \pm ... \pm u_rv_r$ which are characterized by the following property with respect to the basis $\mathbf{e}_1, ..., \mathbf{e}_n$: $(\mathbf{e}_i, \mathbf{e}_j) = 0$ for $i \neq j$, and $(\mathbf{e}_i, \mathbf{e}_i) = 1, -1$ or 0.

We call vectors $\mathbf{u}, \mathbf{v}$ Q-orthogonal if $Q(\mathbf{u}, \mathbf{v}) = 0$. In order to prove the theorem we first construct a basis $\mathbf{f}_1, ..., \mathbf{f}_n$ of pairwise Q-orthogonal vectors. Namely, on the role of $\mathbf{f}_1$ we pick any vector satisfying $Q(\mathbf{f}_1) \neq 0$ (if such a vector does not exist then the quadratic and the corresponding bilinear forms are identically zero and thus any basis $\mathbf{f}_1, ..., \mathbf{f}_n$ will do). Then we consider the subspace V_1 in $\mathbb{R}^n$ of all vectors Q-orthogonal to $\mathbf{f}_1$. It is given by one non-trivial linear equation $Q(\mathbf{f}_1, \mathbf{x}) = 0$ and thus has dimension $n-1$. Next, we restrict the quadratic form to this subspace and repeat our first step: choose $\mathbf{f}_2$ as a vector in V_1 satisfying $Q(\mathbf{f}_2) \neq 0$ (if exists) and pass to the subspace V_2 of all vectors in V_1 which are Q-orthogonal to $\mathbf{f}_2$. The process stops when we have selected $\mathbf{f}_1, ..., \mathbf{f}_r$ in the subspaces $\mathbb{R}^n, V_1, .., V_{r-1}$ of dimensions $n, n-1, ..., n-r+1$, but the next subspace V_r does not contain vectors with $Q(\mathbf{x}) \neq 0$. Of course, it may be due to $r = n$ and V_r being the trivial subspace. If however $r < n$, we complete the construction of the basis by choosing any basis in V_r on the role of $\mathbf{f}_{r+1}, ..., \mathbf{f}_n$. Since each of the vectors $\mathbf{f}_i$ is Q-orthogonal to all previous ones, we have $Q(\mathbf{f}_i, \mathbf{f}_j) = 0$ for $i \neq j$, and also $Q(\mathbf{f}_i, \mathbf{f}_i) = 0$ for $i > r$. In order to complete the proof it remains only to rescale the vectors $\mathbf{f}_1, ..., \mathbf{f}_r$ by the factors $|Q(\mathbf{f}_i, \mathbf{f}_i)|^{-1/2}$ which makes $Q(\mathbf{f}_i, \mathbf{f}_i) = \pm 1$ for $i \leq r$.

Examples. (b) A quadratic form Q in $\mathbb{R}^n$ is called positive if $Q(\mathbf{u}) > 0$ for all $\mathbf{x} \neq \mathbf{0}$. The symmetric bilinear form $Q(\mathbf{u}, \mathbf{v})$ corresponding to a positive quadratic form is called an inner product. The theorem shows that a positive quadratic form takes on $x_1^2 + ... + x_n^2$ in a suitable coordinate system in $\mathbb{R}^n$. Indeed, this is the only one among the functions $\pm x_1^2 \pm ... \pm x_r^2$ which is positive everywhere outside the origin. The proof of the theorem in this case consists in constructing an orthonormal basis of the corresponding inner product $Q(\mathbf{u}, \mathbf{v})$, that is a basis $\mathbf{f}_1, ..., \mathbf{f}_n$ satisfying $Q(\mathbf{f}_i, \mathbf{f}_j) = 0$ for $i \neq j$ and $= 1$ for $i = j$. Thus *any inner product has an orthonormal basis.* This result shows that different choices of a positive symmetric bilinear form on the role of the inner product in $\mathbb{R}^n$ give rise to equivalent Euclidean geometries. Indeed, any inner product in an orthonormal basis takes on the standard form $\langle \mathbf{u}, \mathbf{v} \rangle = u_1 v_1 + ... + u_n v_n$ corresponding to the quadratic form $\langle \mathbf{x}, \mathbf{x} \rangle = x_1^2 + ... + x_n^2$.

(c) Restriction of the quadratic form $x_1^2 + ... + x_n^2$ to a subspace V in $\mathbb{R}^n$ is a positive quadratic form in the subspace. Identifying the subspace with $\mathbb{R}^k$ and applying the theorem we conclude that *any non-zero subspace in the Euclidean space has an orthonormal basis.* For instance, let V be a plane containing given two n-vectors $\mathbf{u}$ and $\mathbf{v}$. If $\mathbf{f}_1, \mathbf{f}_2$ is an orthonormal basis in the plane, then the inner product of vectors in this plane is described by the standard coordinate formula $u_1 v_1 + u_2 v_2$ with respect to this basis. We conclude that *the Schwartz inequality* $\langle \mathbf{u}, \mathbf{v} \rangle^2 \leq \langle \mathbf{u}, \mathbf{u} \rangle \langle \mathbf{v}, \mathbf{v} \rangle$ *holds true for any n-vectors* $\mathbf{u}, \mathbf{v}$. Thus we have just proved the following inequality:

$$(x_1 y_1 + ... + x_n y_n)^2 \leq (x_1^2 + ... + x_n^2)(y_1^2 + ... + y_n^2).$$

The following Proposition shows that the numbers of positive and negative squares in the normal form $\pm X_1^2 \pm ... \pm X_r^2$ do not depend on the choice of the coordinate system but are determined by the quadratic form Q itself.

Proposition. *The numbers p and q, $p+q = r$, of positive and negative squares in $Q = \pm X_1^2 \pm ... \pm X_r^2$ are equal to the maximal dimensions of the subspaces in $\mathbb{R}^n$ where the quadratic form Q (respectively, $-Q$) is positive.*

Proof. The quadratic form $X_1^2 + ... + X_p^2 - X_{p+1}^2 - ... - X_{p+q}^2$ is positive on the p-dimensional subspace $X_{p+1} = ... = X_n = 0$ and non-positive on the subspace W of dimension $n-p$ given by the equations $X_1 = ... = X_p = 0$. Existence of a subspace V of dimension $p+1$ where the quadratic form is positive would lead to a contradiction. Indeed, The subspaces V and W where the quadratic form is positive and non-positive would intersect along a subspace of dimension at least $(p+1)+(n-p)-n=1$ containing therefore non-zero vectors $\mathbf{v}$ with $Q(\mathbf{v})>0$ and $Q(\mathbf{v})\leq 0$.

The maximal dimensions of positive and negative subspaces are called respectively **positive** and **negative inertia indices** of a quadratic form in question. The Proposition and the Theorem together cover the content of the Inertia Theorem formulated in the section 3.1 and show that inertia indices $(p,q), p+q\leq n$, completely characterize those properties of quadratic forms in $\mathbb{R}^n$ which do not depend on the choice of coordinates. The matrix formulation of the Inertia Theorem reads:

Any symmetric matrix Q can be transformed by $Q\mapsto C^tQC$ with invertible C to one and exactly one of the diagonal forms $\begin{bmatrix} I_p & 0 & 0 \\ 0 & -I_q & 0 \\ 0 & 0 & 0 \end{bmatrix}$.

Exercises 3.5.1.

(a) For each of the following quadratic forms $Q(\mathbf{x})$, write down the corresponding symmetric matrix $Q=[q_{ij}]$, the symmetric bilinear form $Q(\mathbf{u},\mathbf{v})$; then, following the proof of the theorem, transform the quadratic form to $\sum \pm x_i^2$ and find the inertia indices: $Q=x_1^2+x_1x_2+x_3x_4$, $Q=x_1x_2+x_2x_3+x_3x_1$, $Q=x_1^2+2x_1x_2+2x_2^2+4x_2x_3+5x_3^2$, $Q=x_1^2-4x_1x_2+2x_1x_3+4x_2^2+x_3^2$.

(b) Let $Q(\mathbf{x})$ be a positive quadratic form. Prove that the determinant $\det[q_{ij}]$ of the corresponding symmetric matrix is positive.

(c) Given a quadratic form $Q(\mathbf{x})$, denote $\Delta_1=q_{11}$, $\Delta_2=q_{11}q_{22}-q_{12}q_{21}$, ..., $\Delta_n=\det[q_{ij}]$ the $k\times k$-minors $\Delta_k=\det[q_{ij}]$, $1\leq i,j\leq k$ of the corresponding symmetric matrix. They are called **principal minors** of the symmetric matrix $[q_{ij}]$. Prove that if Q is positive then $\Delta_1,...,\Delta_n>0$.

(d) Suppose that all principal minors of a quadratic form are non-zero. Following the proof of the theorem, show that the basis $\mathbf{f}_1,...,\mathbf{f}_n$ diagonalizing the quadratic form can be chosen in such a way that $\mathbf{f}_1$ is proportional to $\mathbf{e}_1$, $\mathbf{f}_2$ is a linear combination of $\mathbf{e}_1$ and $\mathbf{e}_2$, $\mathbf{f}_3$ is a linear combination of $\mathbf{e}_1,\mathbf{e}_2,\mathbf{e}_3$, etc.

(e) Deduce that a symmetric matrix Q with non-zero principal minors can be written as $Q=U^tDU$ where U is an invertible upper-triangular matrix, and D is a diagonal matrix with the diagonal entries ± 1.

(f) Deduce the Sylvester theorem: *the negative inertia index q of a quadratic form with non-zero principal minors equals the number of changes of signs in the sequence $\Delta_0=1,\Delta_1,...,\Delta_n$.*

(g) Test the Sylvester theorem in examples of Exercise (a).

(h) Find inertia indices of the quadratic form $\sum_{i\neq j} x_ix_j$.

(i) Classify surfaces in $\mathbb{R}^3$ given by equations $F(x_1,x_2,x_3)=0$, where F is a polynomial of degree ≤ 2, up to linear inhomogeneous changes of coordinates

$$x_i=c_{i1}x_1'+c_{i2}x_2'+c_{i3}x_3'+d_i,\ i=1,2,3.$$

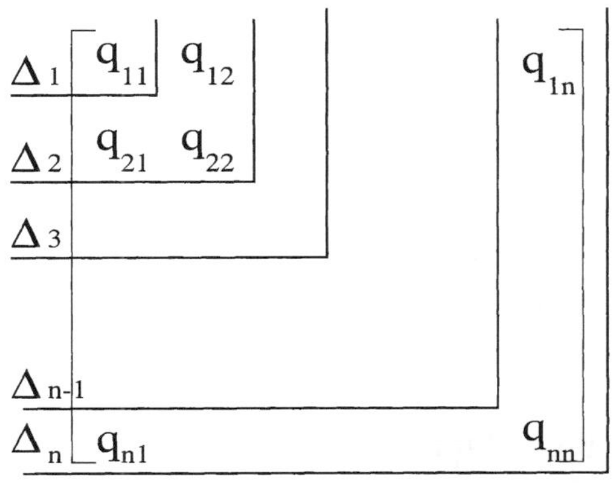

3.5.2. Least square fitting to data. In pure mathematics, there is exactly one straight line through any two distinct points in the plane. In empirical sciences, one tends to suspect a linear dependence $y = \alpha + \beta x$ in any cloud of experimental points graphed on the plane and to explain the deviation of the points from the line by natural experimental errors. The problem we will discuss here is how to find α and β which make the line $y = \alpha + \beta x$ fit the experimental data with minimal error.

More generally, the problem can be formulated as follows. Let $y_1, ..., y_N$ be the values of some function $y = y(x)$ measured experimentally at the points $x_1, ..., x_N$. One assumes that the actual dependence of y on x has the form of a linear combination $f(x) = \alpha_1 f_1(x) + ... + \alpha_n f_n(x)$ of some known functions $f_1, ..., f_n$ where $n < N$ (in the above example $n = 2$, and the known functions are 1 and x). The problem is to *find the coefficients* $\alpha_1, ..., \alpha_n$ *which would minimize the total error*

$$E(\alpha_1, ..., \alpha_n) = (y_1 - f(x_1))^2 + ... + (y_N - f(x_N))^2.$$

We would like to stress that there is no intrinsic reason why one should choose the sum of squares for measuring the error. Any other positive function of $y_i - f(x_i)$, say $E(\alpha) = \sum (y_i - f(x_i))^4$, would seem equally suitable for this role. The unknowns $\alpha_1, ..., \alpha_n$ hidden in the symbols $f(x_i)$ are to be found via minimization of the function $E(\alpha)$. This yields the system of algebraic equations $\partial E/\partial \alpha_1 = 0, ..., \partial E/\partial \alpha_n = 0$ which are non-linear (say, will have degree 3 for the sum of 4-th powers) unless the error function is quadratic. Thus, the choice of the sum of squares is rather dictated by our inability to solve non-linear equations.

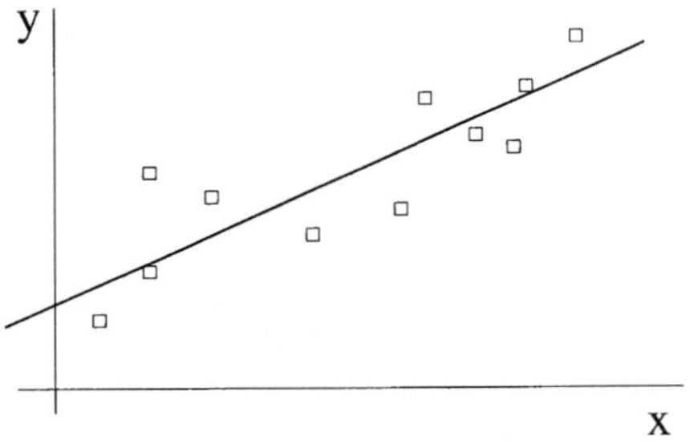

Fitting to data

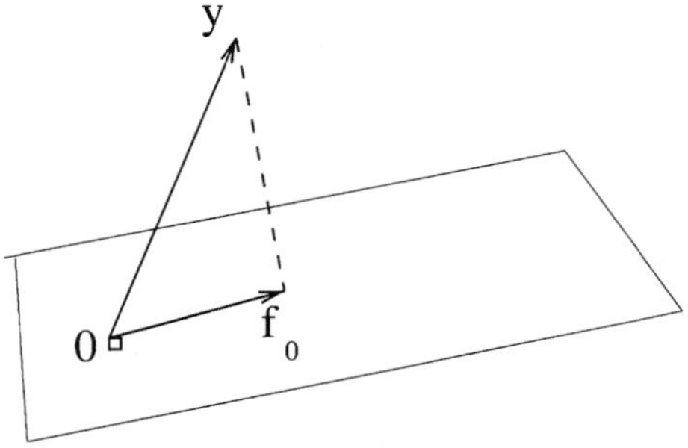

Projection of a vector
to a subspace

We will derive the above system of linear equations for α using some elementary Euclidean geometry instead of multivariable calculus. Let us introduce the $N \times n$-matrix A with the entries $a_{ij} = f_j(x_i)$. The N-vectors $\mathbf{f}_j = (f_j(x_1), ..., f_j(x_N))$ span the column space V of A which therefore consists of all linear combinations $\mathbf{f} = \alpha_1\mathbf{f}_1 + ... + \alpha_n\mathbf{f}_n$. Let $\mathbf{y} = (y_1, ..., y_N)$ be the N-vector of experimental data. Minimization of the error function $E = |\mathbf{y} - \mathbf{f}|^2$ can be now interpreted as *finding the point* $\mathbf{f}_0$ *in the subspace* V *which minimizes the Euclidean distance to* $\mathbf{y}$. We will show that *there exists a unique point* $\mathbf{f}_0$ *in* V *which minimizes the distance to* $\mathbf{y}$*, and that* $\mathbf{f}_0$ *is characterized by the condition that the vector* $\mathbf{y} - \mathbf{f}_0$ *is orthogonal to* V*:* $\langle \mathbf{y} - \mathbf{f}_0, \mathbf{f}\rangle = 0$ *for any* $\mathbf{f}$ *from* V. One calls this $\mathbf{f}_0$ the **orthogonal projection** of $\mathbf{y}$ to the subspace V.

Indeed, if the orthogonality condition is satisfied then $|\mathbf{y}-\mathbf{f}|^2 = |\mathbf{y}-\mathbf{f}_0|^2 + |\mathbf{f}_0 - \mathbf{f}|^2$ by the Pythagorean theorem and hence $|\mathbf{y} - \mathbf{f}|^2 > |\mathbf{y} - \mathbf{f}_0|^2$ unless $\mathbf{f} = \mathbf{f}_0$. This proves uniqueness. In order to prove existence, let us pick an orthonormal basis $\mathbf{e}_1, ..., \mathbf{e}_m$ in V and put

$$\mathbf{f}_0 = \langle \mathbf{y}, \mathbf{e}_1\rangle \mathbf{e}_1 + ... + \langle \mathbf{y}, \mathbf{e}_m\rangle \mathbf{e}_m.$$

It is immediate to check that $\langle \mathbf{y} - \mathbf{f}_0, \mathbf{e}_j\rangle = 0$ for all $j = 1, ..., m$. Thus $\mathbf{y} - \mathbf{f}_0$ is orthogonal to any linear combination of $\mathbf{e}_1, ..., \mathbf{e}_m$.

As a by-product, we get a nice formula describing the orthogonal projection $\mathbf{f}_0$ of a vector $\mathbf{y}$ to a subspace in terms of an orthonormal basis $\mathbf{e}_1, ..., \mathbf{e}_m$ in the subspace.

The orthogonality condition $\langle \mathbf{f}, \mathbf{y} - \mathbf{f}_0\rangle = 0$ is sufficient to check for $\mathbf{f} = \mathbf{f}_1, ..., \mathbf{f}_n$ spanning V. Thus, we arrive to the system of n linear equations for n coefficients of the linear combination $\mathbf{f}_0 = \alpha_1\mathbf{f}_1 + ... + \alpha_n\mathbf{f}_n$:

$$\langle \mathbf{f}_i, \mathbf{f}_1\rangle \alpha_1 + ... + \langle \mathbf{f}_i, \mathbf{f}_n\rangle \alpha_n = \langle \mathbf{f}_i, \mathbf{y}\rangle, \ i = 1, ..., n.$$

If the vectors $\mathbf{f}_1, ..., \mathbf{f}_n$ are linearly independent (that is $n = m$), the solution to this system is unique. In the matrix form the system can be written as $A^tA\mathbf{a} = A^t\mathbf{y}$ where $\mathbf{a} = (\alpha_1, ..., \alpha_n)$. Solving the system one finds the solution to our problem of least square fitting to the experimental data.

Example. If $n = 2$ and $f_1 = 1, f_2 = x$, we have $\mathbf{f}_1 = (1, ..., 1)$, $\mathbf{f}_2 = (x_1, ..., x_N)$, $\mathbf{y} = (y_1, ..., y_N)$. The linear 2×2-system for α, β reads

$$\alpha N + \beta \sum x_i = \sum y_i, \quad \alpha \sum x_i + \beta \sum x_i^2 = \sum x_i y_i.$$

Thus the line "passing through" the points $(x_1, y_1), ..., (x_N, y_N)$ with the least square error is given by the equation $y = \alpha + \beta x$ where

$$\alpha = \frac{(\sum y_i)(\sum x_i^2) - (\sum x_i y_i)(\sum x_i)}{N \sum x_i^2 - (\sum x_i)^2}, \quad \beta = \frac{N \sum x_i y_i - (\sum x_i)(\sum y_i)}{N \sum x_i^2 - (\sum x_i)^2}.$$

Exercises 3.5.2.

(a) Find the linear function $y = \alpha x + \beta$ which provides the least square fitting to the data

$$y = 1, 4, 9, 16$$

$$x = 1, 2, 3, 4.$$

(b) Find the quadratic function $y = \alpha x^2 + \beta x + \gamma$ which provides the least square fitting to the data

$$y = -1, 1, -1, 1$$

$$x = 0, 1, 2, 3.$$

(c) Find the projection of the vector $\mathbf{x} = (1,1,1,1,1)$ to the plane spanned by $(1,0,-1,0,1)$ and $(0,1,0,-1,0)$ and compute the distance from $\mathbf{x}$ to the plane.

(d) *Gram – Schmidt orthogonalization.*

Given a subspace V in the Euclidean n-space and a basis $\mathbf{f}_1, ..., \mathbf{f}_k$ in the subspace, one constructs an orthonormal basis $\mathbf{u}_1, ..., \mathbf{u}_k$ in V as follows: Normalize $\mathbf{f}_1$ to the unit length and denote the result $\mathbf{u}_1$. Subtract from $\mathbf{f}_2$ its projection to the line spanned by $\mathbf{u}_1$ and normalize the difference to the unit length. Denote the result $\mathbf{u}_2$. Subtract from $\mathbf{f}_3$ its projection to the plane spanned by $\mathbf{u}_1$ and $\mathbf{u}_2$ and normalize the difference to the unit length. Denote the result $\mathbf{u}_3$. And so on ...

Apply the algorithm to the subspace in $\mathbb{R}^4$ spanned by

$$\mathbf{f}_1 = (1,1,1,1), \quad \mathbf{f}_2 = (1,-1,-1,1) \quad \mathbf{f}_3 = (3,1,1,-1).$$

Prove that $\mathbf{u}_1, ..., \mathbf{u}_k$ form an orthonormal basis in V satisfying the condition that $\mathbf{u}_i$ is a linear combination of $\mathbf{f}_1, ..., \mathbf{f}_i$.

3.5.3. Orthonormal bases. Our next goal is to classify pairs of quadratic forms in n variables up to linear changes of the variables, assuming that one of the quadratic forms is positive. According to the Inertia Theorem, there exists a linear change of coordinates which transforms the positive quadratic form to $x_1^2 + ... + x_n^2$. The same change of coordinates applied to the second quadratic form yields some quadratic form $Q(\mathbf{x}) = \sum_{i,j} q_{ij} x_i x_j$. Our problem is therefore equivalent to the classification of quadratic forms $Q(\mathbf{x})$ up to linear transformations *preserving Euclidean inner squares* $\langle \mathbf{x}, \mathbf{x} \rangle = x_1^2 + ... + x_n^2$ *of all vectors.* Furthermore, the Euclidean inner product $\langle \mathbf{x}, \mathbf{y} \rangle = x_1 y_1 + ... + x_n y_n$ can be expressed via addition of vectors and their lengths as $\langle \mathbf{x}, \mathbf{y} \rangle = (\langle \mathbf{x}+\mathbf{y}, \mathbf{x}+\mathbf{y} \rangle - \langle \mathbf{x}, \mathbf{x} \rangle - \langle \mathbf{y}, \mathbf{y} \rangle)/2$. Since linear transformations send sums of vectors to sums, we conclude that linear transformations preserving lengths of all vectors actually preserve inner products of all vectors. Linear transformations U from $\mathbb{R}^n$ to itself which preserve inner products, $\langle U\mathbf{x}, U\mathbf{y} \rangle = \langle \mathbf{x}, \mathbf{y} \rangle$ for all $\mathbf{x}$ and $\mathbf{y}$, are called orthogonal transformations. Thus, our problem is to *classify quadratic forms in* $\mathbb{R}^n$ *up to orthogonal transformations.* We begin with the following proposition which gives several equivalent characterizations of orthogonal transformations.

Proposition. *The following properties of a linear transformation U are equivalent:*

(a) U is orthogonal.
(b) The matrix of U in an orthonormal basis satisfies $U^t U = I$.
(c) $U^{-1} = U^t$.
(d) The matrix of U in an orthonormal basis satisfies $UU^t = I$.
(e) U^t is orthogonal.
(f) Columns of the matrix form an orthonormal basis.
(g) Rows of the matrix form an orthonormal basis.
(h) U transforms orthonormal bases to orthonormal bases.
(i) U transforms an orthonormal basis to an orthonormal basis.

Proof. If U is orthogonal then $\langle \mathbf{x}, \mathbf{y} \rangle = \langle U\mathbf{x}, U\mathbf{y} \rangle = \langle \mathbf{x}, U^t U\mathbf{y} \rangle$ for all $\mathbf{x}, \mathbf{y}$ which is possible only if $U^t U = I$. If $U^t U = I$ then $\det U = \pm 1$ and in particular U is invertible and $U^{-1} = U^t$. If U is invertible and $U^{-1} = U^t$ then $UU^t = I$. If $UU^t = I$ then $\langle U^t\mathbf{x}, U^t\mathbf{y} \rangle = \langle \mathbf{x}, UU^t\mathbf{y} \rangle = \langle \mathbf{x}, \mathbf{y} \rangle$ for all $\mathbf{x}, \mathbf{y}$ and hence U^t is orthogonal. If U^t is orthogonal, then the previous chain of implications shows that $U^{tt} = U$ is orthogonal and therefore proves the equivalence of (a),(b),(c),(d),(e).

Obviously, the equalities $U^tU = I$ and $UU^t = I$ are equivalent to the properties of columns and respectively — rows of the matrix U to form an orthonormal basis Thus (f), (g) are equivalent to (b), (d).

If U is orthogonal than it transforms orthonormal bases to orthonormal bases (since it is invertible, as we already know, and preserves inner products of all vectors). Thus (a) implies (h). Obviously (h) implies (i). On the other hand (i) means that the matrix of U in some orthonormal basis satisfies $U^tU = I$ and hence (i) implies (b) and hence (a).

Examples. (a) As we know from Chapter 1, orthogonal transformations in $\mathbb{R}^2$ are rotations about the origin or reflections about a line through the origin.

(b) Rotation in the space about a line through the origin are examples of orthogonal transformations in $\mathbb{R}^3$. Let $\mathbf{f}_1$ be one of (the two) unit vectors in the line, and $\mathbf{f}_2, \mathbf{f}_3$ be an orthonormal basis in the plane perpendicular to this line. Composing the rotation with the reflection $\mathbf{f}_1 \mapsto -\mathbf{f}_1$, $\mathbf{f}_2 \mapsto \mathbf{f}_2$, $\mathbf{f}_3 \mapsto \mathbf{f}_3$ about the plane perpendicular to the axis of rotation we get another example of orthogonal transformations in $\mathbb{R}^3$. The rotation about $\mathbf{f}_1$ through the angle θ and its composition with the reflection have in the basis $\mathbf{f}_1, \mathbf{f}_2, \mathbf{f}_3$ the following matrices:

$$\begin{bmatrix} 1 & 0 & 0 \\ 0 & \cos\theta & -\sin\theta \\ 0 & \sin\theta & \cos\theta \end{bmatrix}, \quad \begin{bmatrix} -1 & 0 & 0 \\ 0 & \cos\theta & -\sin\theta \\ 0 & \sin\theta & \cos\theta \end{bmatrix}.$$

The matrices have determinants 1 and -1 respectively. Notice that these two types of orthogonal transformations in $\mathbb{R}^3$ depend on 3 parameters: 2 angles are needed in order to specify the direction of the line spanned by $\mathbf{f}_1$, and the third parameter is the angle θ of rotation. This dimension counting agrees with the fact that orthonormal bases in $\mathbb{R}^3$ are singled out by 6 equations on 9 coordinates of the three basic vectors (pairwise orthogonality and unit lengths of the vectors).

In fact *any orthogonal transformation in $\mathbb{R}^3$ is a rotation about a line or the composition of such a rotation with the reflection about the plane perpendicular to this line.* Indeed, consider the characteristic polynomial $\det(\lambda I_3 - U)$ of a 3×3-matrix U. It has degree 3 and hence — at least one real root λ_0. The system $U\mathbf{x} = \lambda_0\mathbf{x}$ has a non-trivial solution $\mathbf{x}_0$. If U is orthogonal then $\langle \mathbf{x}_0, \mathbf{x}_0\rangle = \langle U\mathbf{x}_0, U\mathbf{x}_0\rangle = \lambda_0^2\langle \mathbf{x}_0, \mathbf{x}_0\rangle$ and hence $\lambda_0 = \pm 1$. Normalizing $\mathbf{x}_0$ to the unit length we get a unit vector $\mathbf{f}_1$ such that $U\mathbf{f}_1 = \pm\mathbf{f}_1$. Since U preserves inner products and preserves the line spanned by $\mathbf{f}_1$ it also preserves the plane perpendicular to $\mathbf{f}_1$ and acts on this plane as an orthogonal transformation in $\mathbb{R}^2$. Assuming that U acts in the plane as a rotation, we pick any orthonormal basis $\mathbf{f}_2, \mathbf{f}_3$ in this plane and find that U has one of the above matrices in the orthonormal basis $\mathbf{f}_1, \mathbf{f}_2, \mathbf{f}_3$. If U acts in the plane as a reflection, that is $\mathbf{f}_2 \mapsto \mathbf{f}_2$, $\mathbf{f}_3 \mapsto -\mathbf{f}_3$, then U is also described by one of the above matrices in a reodered basis (figure out yourself — which one?)

In the above argument, we considered the plane perpendicular to a given line in $\mathbb{R}^3$. This construction is an example of the orthogonal complement to a subspace in the Euclidean n-space. Let V be a subspace in $\mathbb{R}^n$ of dimension l. A vector $\mathbf{w}$ is called **orthogonal to** V if it is orthogonal to any vector from V. If $\mathbf{u}$ and $\mathbf{w}$ are orthogonal to V then their linear combinations are also orthogonal to V: for any $\mathbf{v}$ from V

$$\langle \alpha\mathbf{u} + \beta\mathbf{w}, \mathbf{v}\rangle = \alpha\langle \mathbf{u}, \mathbf{v}\rangle + \beta\langle \mathbf{w}, \mathbf{v}\rangle = \alpha 0 + \beta 0 = 0.$$

Thus vectors orthogonal to V form a subspace W which is called the **orthogonal complement to** V. The subspaces V and W intersect only at the origin. Indeed, a common vector $\mathbf{u}$ of V and W is orthogonal to itself, $\langle \mathbf{u}, \mathbf{u} \rangle = 0$, and thus must be equal to $\mathbf{0}$. Thus the dimension of W does not exceed $n - l$. On the other hand, W is given by l linear equations $\langle \mathbf{w}, \mathbf{v}_1 \rangle = ... = \langle \mathbf{w}, \mathbf{v}_l \rangle = 0$ where $\mathbf{v}_1, ..., \mathbf{v}_l$ is a basis in V and therefore must have dimension $n-l$ at least. Thus $\dim V + \dim W = n$. Also, all vectors from V are orthogonal to W and hence V is the orthogonal complement to W. In particular, if $\mathbf{v}_1, ..., \mathbf{v}_l$ and $\mathbf{w}_1, ..., \mathbf{w}_{n-l}$ are orthonormal bases in V and W then their union is an orthonormal basis in $\mathbb{R}^n$.

Example. (c) An $m \times n$-matrix A and its transpose A^t define linear operators of the same rank r from $\mathbb{R}^n$ to $\mathbb{R}^m$ and from $\mathbb{R}^m$ to $\mathbb{R}^n$ respectively. *The range (= the column space) of A is the orthogonal complement in $\mathbb{R}^m$ to the null-space of A^t, and the range of A^t (= the row space of A) is the orthogonal complement to the null-space of A in $\mathbb{R}^n$.* Indeed, if $\mathbf{y} = A\mathbf{x}$ and $A^t\mathbf{y}' = \mathbf{0}$ then

$$\langle \mathbf{y}, \mathbf{y}' \rangle = \langle A\mathbf{x}, \mathbf{y}' \rangle = \langle \mathbf{x}, A^t\mathbf{y}' \rangle = \langle \mathbf{x}, \mathbf{0} \rangle = 0.$$

Similarly, all vectors from the range of A^t are orthogonal to all vectors from the null-space of A. The rest follows from dimension counting.

Exercises 3.5.3.

(a) Find the matrix (in the standard basis $\mathbf{e}_1, \mathbf{e}_2, \mathbf{e}_3$) of the rotation through the angle $\pm\theta$ about the line in $\mathbb{R}^3$ spanned by the vector $(1, 1, 1)$.

(b) Find the place in the above description of orthogonal transformations in $\mathbb{R}^3$ for (i) a reflection about a plane, (ii) reflection about a line, (iii) composition of reflections about two different planes, (iv) composition of three such reflections.

(c) Show that real eigenvalues of an orthogonal matrix are equal to ± 1.

Give an example of an orthogonal 4×4 matrix with no real eigenvalues.

(d) Find the orthogonal complement to the plane in $\mathbb{R}^4$ spanned by the vectors $(1, 1, 1, 1)$ and $(-1, 1, -1, 1)$. Find the orthogonal complement to the plane in $\mathbb{R}^4$ given by the equations $x_1 + x_2 + x_3 = 0$, $x_2 - x_3 + x_4 = 0$. Find an orthonormal bases in these orthogonal complements.

(e) *Cayley transform.* Let U be an orthogonal matrix without eigenvalue -1. Prove that the matrix $\Omega = (I - U)(I + U)^{-1}$ is anti-symmetric. Let Ω be an anti-symmetric matrix. Prove that the matrix $U = (I - \Omega)(I + \Omega)^{-1}$ is orthogonal.

(f) *LU-factorization.* Let $\mathbf{u}_1, ..., \mathbf{u}_n$ be the orthonormal basis in $\mathbb{R}^n$ obtained by the Gram-Schmidt orthogonalization of a basis $\mathbf{f}_1, ..., \mathbf{f}_n$ (see Exercise 3.5.2(*d*)). Represent the property of $\mathbf{u}_i$ to be linear combinations of $\mathbf{f}_1, ..., \mathbf{f}_i$ in a matrix form and deduce that *any invertible matrix can be written as the product UR of an orthogonal matrix U and an upper - triangular matrix R.*

Applying transposition show that, similarly, *any invertible matrix is the product LU of a lower-triangular and an orthogonal matrix.*

3.5.4. Orthogonal diagonalization. As we have already mentioned in Section 3.5.1, the Orthogonal Diagonalization Theorem formulated in Section 3.1 is reduced to the existence of an orthogonal transformation U in $\mathbb{R}^n$ which transforms a given quadratic form $Q(x)$ to $\lambda_1 x_1^2 + ... + \lambda_n x_n^2$. In the matrix form, $U^t Q U = \Lambda$ where Λ is to be diagonal. Since $U^t = U^{-1}$, this means $QU = U\Lambda$, that is columns of U are to form an orthonormal basis of eigenvectors of Q with the eigenvalues $\lambda_1, ..., \lambda_n$. Thus the Orthogonal Diagonalization Theorem can be reformulated this way:

Theorem. *Any symmetric matrix Q has an orthonormal basis of eigenvectors.*

Proof. First, we want to show that *eigenvalues of a symmetric matrix are real.* For this, let λ_0 be a complex root of the characteristic polynomial $\det(\lambda I - Q)$ (λ_0 exists by the Fundamental Theorem of Algebra). Then the system of linear equations $Q\mathbf{z} = \lambda_0\mathbf{z}$ has a non-trivial complex solution $\mathbf{z} = (z_1, ..., z_n) \neq (0, ..., 0)$. We put $\bar{\mathbf{z}} = (\bar{z}_1, ..., \bar{z}_n)$, note that $Q\bar{\mathbf{z}} = \bar{\lambda}_0\bar{\mathbf{z}}$ (since Q is real) and compute the number $\langle Q\mathbf{z}, \bar{\mathbf{z}}\rangle = \sum q_{ij} z_j \bar{z}_i$ in two ways (using $Q = Q^t$):

$$\lambda_0 \langle \mathbf{z}, \bar{\mathbf{z}}\rangle = \langle Q\mathbf{z}, \bar{\mathbf{z}}\rangle = \langle \mathbf{z}, Q\bar{\mathbf{z}}\rangle = \bar{\lambda}_0 \langle \mathbf{z}, \bar{\mathbf{z}}\rangle.$$

Since $\langle \mathbf{z}, \bar{\mathbf{z}}\rangle = |z_1|^2 + ... + |z_n|^2 > 0$, we conclude that $\lambda_0 = \bar{\lambda}_0$ meaning that λ_0 is real indeed.

Second, we claim that *eigenvectors of a symmetric matrix corresponding to different eigenvalues are orthogonal to each other.* Indeed, let $Q\mathbf{u} = \lambda\mathbf{u}$, $Q\mathbf{v} = \mu\mathbf{v}$, $\mathbf{u} \neq \mathbf{0}$, $\mathbf{v} \neq \mathbf{0}$ and $Q^t = Q$. Then

$$\lambda\langle \mathbf{u}, \mathbf{v}\rangle = \langle Q\mathbf{u}, \mathbf{v}\rangle = \langle \mathbf{u}, Q\mathbf{v}\rangle = \mu\langle \mathbf{u}, \mathbf{v}\rangle$$

and hence $\langle \mathbf{u}, \mathbf{v}\rangle = 0$ unless $\lambda = \mu$.

An orthonormal basis of eigenvectors of Q can be constructed as follows. Let $\lambda_1, ..., \lambda_r$ denote the set of distinct roots of the characteristic polynomial, and $V_1, ..., V_r$ denote the corresponding eigenspaces. We pick orthonormal bases in the subspaces V_i and consider the resulting set of vectors $\mathbf{v}_1, ..., \mathbf{v}_k$. It follows from the previous paragraph that the set is orthonormal (and hence — linearly independent). We claim that in fact the vectors span $\mathbb{R}^n$ so that $k = n$ and the vectors form a *basis* in $\mathbb{R}^n$. In order to justify the claim we need the following

Lemma. If all vectors from a subspace V are transformed by Q to vectors from the same subspace V, then the same is true about the orthogonal complement W of V.

Indeed, if $\mathbf{w}$ is orthogonal to all vectors from V, then $\langle Q\mathbf{w}, \mathbf{v}\rangle = \langle \mathbf{w}, Q\mathbf{v}\rangle = 0$ for all $\mathbf{v}$ from V (since $Q\mathbf{v}$ is in V) and hence $Q\mathbf{w}$ is in W.

Finally, we apply the lemma to the subspace $V = Span(\mathbf{v}_1, ..., \mathbf{v}_k)$ containing *all* eigenvectors of Q and show that the orthogonal complement W must contain eigenvectors of Q too (in contradiction with the choice of V) unless $W = \mathbf{0}$. Namely, if $W \neq \mathbf{0}$, consider the linear transformation from W to itself defined by Q. In an orthonormal basis of W, the matrix of this linear transformation will be symmetric, since the property $\langle Q\mathbf{w}, \mathbf{w}'\rangle = \langle \mathbf{w}, Q\mathbf{w}'\rangle$ holds true for any $\mathbf{w}, \mathbf{w}'$ from W. Thus our linear transformation in W has real eigenvalues (according to the first step of the proof) and hence — real eigenvectors. This contradiction shows that $W = \mathbf{0}$ and hence $V = \mathbb{R}^n$.

Example. (a) The matrix

$$Q = \begin{bmatrix} 2 & -1 & 0 & -1 \\ -1 & 2 & -1 & 0 \\ 0 & -1 & 2 & -1 \\ -1 & 0 & -1 & 2 \end{bmatrix}$$

of the quadratic form

$$Q(\mathbf{x}) = (x_1 - x_2)^2 + (x_2 - x_3)^2 + (x_3 - x_4)^2 + (x_4 - x_1)^2$$

has the characteristic polynomial $\lambda^4 - 8\lambda^3 + 20\lambda^2 - 16\lambda = \lambda(\lambda - 2)^2(\lambda - 4)$. The columns of the matrix

$$U' = \begin{bmatrix} 1 & 1 & 1 & 1 \\ 1 & 1 & -1 & -1 \\ 1 & -1 & -1 & 1 \\ 1 & -1 & 1 & -1 \end{bmatrix}$$

(some vertices of the $4D$-cube studied in Section 3.3.1) are mutually perpendicular eigenvectors of Q with the eigenvalues 0, 2, 2, 4 respectively. When divided by 2 they form an orthonormal basis in $\mathbb{R}^4$ (and the corresponding matrix $U = U'/2$ is orthogonal). The change of variables $\mathbf{x} = U\mathbf{y}$ transforms $Q(\mathbf{x})$ to $0y_1^2 + 2y_1^2 + 2y_3^2 + 4y_4^2$.

Exercises 3.5.4.

(a) Following the proof of the Orthogonal Diagonalization Theorem find an orthonormal basis diagonalizing the symmetric matrix

$$\begin{bmatrix} 1 & 1 & 1 \\ 1 & 1 & 1 \\ 1 & 1 & 1 \end{bmatrix}.$$

(b) Transform the following quadratic forms to the form $\sum \lambda_i x_i^2$ by orthogonal transformations:

$$2x_1^2 + x_2^2 - 4x_1x_2 - 4x_2x_3, \quad x_1^2 + 2x_2^2 + 3x_3^2 - 4x_1x_2 - 4x_2x_3, \quad 3x_1^2 + 4x_2^2 + 5x_3^2 + 4x_1x_2 - 4x_2x_3,$$

$$x_1^2 - 2x_2^2 - 2x_3^2 - 4x_1x_2 + 4x_1x_3 + 8x_2x_3, \quad x_1^2 + x_2^2 + x_3^2 + x_4^2 + 2x_1x_2 - 2x_1x_4 - 2x_2x_3 + 2x_3x_4.$$

(c) Three knights at The King Arthur's round table are served unequal amounts of cereal. At the moment n each knight borrows one half of cereal from the plates of his left and right neighbors. Find the distribution of cereal in the plates at the moment $n = 1999$ when The King Arthur joins the party.

(d) Show that maximum (minimum) of the quadratic form $\lambda_1 x_1^2 + ... + \lambda_n x_n^2$ restricted to the unit sphere $x_1^2 + ... + x_n^2$ equals the maximal (minimal) number among $\lambda_1, ..., \lambda_n$.

(e) The hypersurface in $\mathbb{R}^n$ given by the equation

$$\frac{x_1^2}{a_1^2} + ... + \frac{x_n^2}{a_n^2} = 1, \quad a_1 \geq a_2 \geq ... \geq a_n,$$

is called the normal ellipsoid with semiaxes $a_1, ..., a_n$. Show that for any positive quadratic form $Q(\mathbf{x})$, the ellipsoid defined by the equation $Q(\mathbf{x}) = 1$ is obtained from one of the normal ellipsoids by an orthogonal transformation.

(f) Show that the intersection of an ellipsoid E in $\mathbb{R}^n$ with a linear subspace is an ellipsoid E' in this subspace. Prove that the largest semiaxes a_1 and a_1' of E and E' satisfy $a_1' \leq a_1$.

(g) Classify quadratic surfaces

$$Q(x_1, x_2, x_3) = 1$$

in $\mathbb{R}^3$ up to orthogonal changes of coordinates (here Q is a quadratic form). Sketch the corresponding pictures.

3.5.5. Small oscillations. Let us consider the system of 4 identical masses m positioned at the vertices of a square which are cyclicly connected by 4 identical elastic springs and can oscillate in the direction perpendicular to the plane of the square. [1] Assuming that the oscillations are small, we can describe motion of the masses by solutions to the following system of Newton equations (mass $\times$ acceleration = force):

$$\begin{aligned} m\ddot{x}_1 &= -k(x_1 - x_4) - k(x_1 - x_2), \\ m\ddot{x}_2 &= -k(x_2 - x_1) - k(x_2 - x_3), \\ m\ddot{x}_3 &= -k(x_3 - x_2) - k(x_3 - x_4), \\ m\ddot{x}_4 &= -k(x_4 - x_3) - k(x_4 - x_1). \end{aligned}$$

[1] We may assume that the springs are stretched, but the masses are confined on vertical rods and can only slide along them without friction. When a spring of length L is horizontal ($\Delta x = 0$), the stretching force T is compensated by the reactions of the rods. When $\Delta x \neq 0$, the horizontal component of the stretching force is still compensated, but the vertical component contributes to the RHS of the Newton equations. When Δx is small, the contribution equals approximately $-T(\Delta x)/L$ (so that $k = -T/L$.)

Here $x_1, ..., x_4$ are the coordinates of the masses in the direction perpendicular to the plane, and k characterizes rigidity of the springs. In fact the ODE system can be read off a pair of quadratic forms — the kinetic energy

$$K(\dot{\mathbf{x}}) = \frac{m\dot{x}_1^2}{2} + \frac{m\dot{x}_2^2}{2} + \frac{m\dot{x}_3^2}{2} + \frac{m\dot{x}_4^2}{2},$$

and the potential energy

$$P(\mathbf{x}) = k\frac{(x_1 - x_2)^2}{2} + k\frac{(x_2 - x_3)^2}{2} + k\frac{(x_3 - x_4)^2}{2} + k\frac{(x_4 - x_1)^2}{2}.$$

Namely, for any conservative mechanical system with quadratic kinetic and potential energy functions

$$K(\dot{\mathbf{x}}) = \frac{1}{2}\langle M\dot{\mathbf{x}}, \dot{\mathbf{x}}\rangle, \; P(\mathbf{x}) = \frac{1}{2}\langle Q\mathbf{x}, \mathbf{x}\rangle$$

the equations of motion take on the form

$$M\ddot{\mathbf{x}} = -Q\mathbf{x}.$$

A linear change of variables $\mathbf{x} = C\mathbf{y}$ transforms the kinetic and potential energy functions to a new form with the matrices $M' = C^tMC$ and $Q' = C^tQC$. On the other hand, the same change of variables transforms the ODE system $M\ddot{\mathbf{x}} = -Q\mathbf{x}$ to $MC\ddot{\mathbf{y}} = -QC\mathbf{y}$. Multiplying by C^t we get $M'\ddot{\mathbf{y}} = -Q'\mathbf{y}$ and see that the relationship between K, P and the ODE system is preserved. The relationship is therefore intrinsic, i. e. independent on the choice of coordinates.

Since the kinetic energy is positive we can apply the Orthogonal Diagonalization Theorem (in the form described in Section 3.1) in order to transform K and P simultaneously to

$$\frac{1}{2}(\dot{y}_1^2 + ... + \dot{y}_n^2), \text{ and } \frac{1}{2}(\lambda_1 y_1^2 + ... + \lambda_n y_n^2).$$

The corresponding ODE system splits into unlinked 2-nd order ODEs

$$\ddot{y}_1 = -\lambda_1 y_1, \; ..., \; \ddot{y}_n = -\lambda_n y_n.$$

When the potential energy is also positive, we obtain a system of n unlinked harmonic oscillators with the frequencies $\omega = \sqrt{\lambda_1}, ..., \sqrt{\lambda_n}$.

Let us examine our system of 4 masses tied by the springs. The kinetic energy function is proportional to the square length of the velocity vector $\dot{\mathbf{x}}$. The potential energy function is proportional to the quadratic form studied in Example (a) of the previous section. Thus we can use the eigenvectors and eigenvalues found there. In the coordinate system (y_1, y_2, y_3, y_4) the ODE system reads $m\ddot{y}_1 = 0$, $m\ddot{y}_2 = -2ky_2$, $m\ddot{y}_3 = -2ky_3$, $m\ddot{y}_4 = -4ky_4$ and is easy to solve explicitly.

The first eigenvector $(x_1, x_2, x_3, x_4) = (1, 1, 1, 1)$ describes the simultaneous shift of the 4 masses in the direction perpendicular to the plane. Since the corresponding eigenvalue $\lambda_1 = 0$, the solution $y_1(t) = y_1(0) + \dot{y}_1(0)t$ does not have oscillating character and represents the "free particle" motion of the mass-spring system as a whole.

The next eigenvector $(1, 1, -1, -1)$ corresponds to the mode of oscillation with the frequency $\sqrt{2k/m}$ where the 1-st and 2-nd masses as well as the 3-rd and 4-th masses move together, but the pairs move in the opposite directions.

Similarly, the eigenvector $(1, -1, -1, 1)$ corresponds to the oscillation mode when the masses $2, 3$ and $1, 4$ move in pairs.

The last eigenvector $(1, -1, 1, -1)$ corresponds to the diagonal pairs $1, 3$ and $2, 4$ moving together. The oscillation frequency $\sqrt{4k/m}$ here is greater than in the previous two cases.

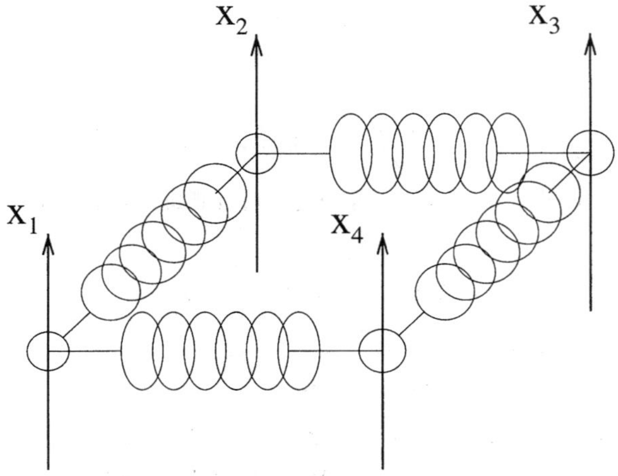

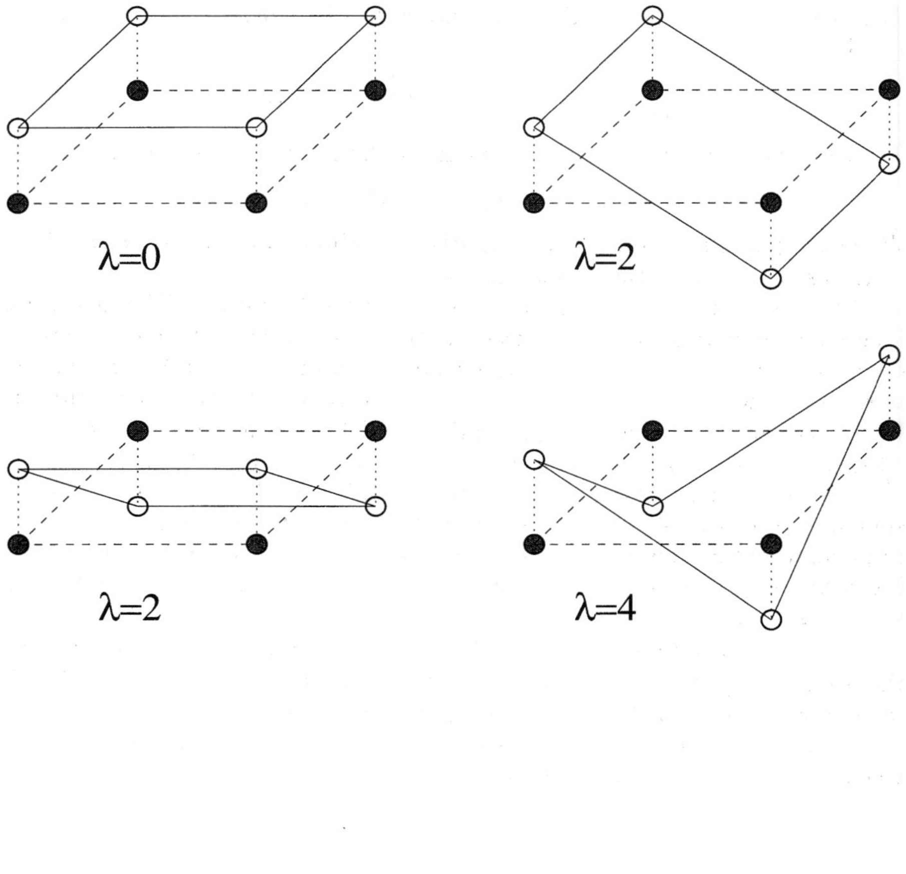

A general motion of the configuration point (x_1, x_2, x_3, x_4) in the configuration space [2] $\mathbb{R}^4$ can be understood as a superposition of the above four modes. Even though the general motion may look complicated in the original coordinate system (related to the coordinate $4D$-cube), we see that it actually consists of four simple independent motions in the directions of the eigenvectors (the axes of the $4D$-"octahedron" formed by "black" vertices of the $4D$-cube shown on the picture in Section 3.3.1).

Similarly to the above example, the Orthogonal Diagonalization Theorem guarantees that *small oscillations in any conservative mechanical system near a local minimum of potential energy are described as superpositions of independent harmonic oscillations.*

Exercises 3.5.5.

(a) Using Exercise 3.5.4(f) explain why a broken bell sounds lower than a new one.

(b) Find frequencies and describe the modes of oscillations in the mass-spring system of 3 identical masses positioned at the vertices of the regular triangle. The same — for 6 masses positioned at the vertices of the regular hexagon (like the 6 carbon atoms in benzene molecules). The same — in the case of the regular n-gon for arbitrary n.

[2] The phase space of our system has dimension 8 since a phase point is to represent not only the configuration vector $\mathbf{x}$ but also the velocity vector $\dot{\mathbf{x}}$.

3.6. Eigenvectors

Classification of linear transformations from a space to itself up to change of coordinates and applications to linear constant coefficients ODE systems is the subject of this section.

3.6.1. Diagonalization theorem. Let $\mathbf{x}' = A\mathbf{x}$ be a linear transformation from $\mathbb{R}^n$ to itself. The $n \times n$-matrix A of the linear transformation depends on the choice of the coordinate system and is subject to a similarity transformation $A \mapsto C^{-1}AC$ when the coordinate system changes: if $\mathbf{x} = C\mathbf{y}$ (and respectively $\mathbf{x}' = C\mathbf{y}'$) we have $\mathbf{y}' = C^{-1}\mathbf{x}' = C^{-1}A\mathbf{x} = C^{-1}AC\mathbf{y}$. As we remarked in Section 3.1, the classification of matrices up to similarity transformations looks simpler if we allow complex matrices on the role of both A and C. So, we introduce the space $\mathbb{C}^n$ of all complex n-vectors, that is columns $(z_1, ..., z_n)^t$ of complex numbers [3] with componentwise addition and multiplication by (complex) scalars. A linear transformation $\mathbf{z}' = A\mathbf{z}$ from $\mathbb{C}^n$ to itself is described by the matrix multiplication with the coefficient matrix $A = [a_{ij}]$ where the matrix entries a_{ij} are allowed to be complex numbers. The characteristic polynomial $\det(\lambda I - A)$ of the matrix A is therefore a degree n polynomial in λ which has complex coefficients. According to the Fundamental Theorem of Algebra it has n complex roots $\lambda_1, ..., \lambda_n$, possibly — multiple. We will assume for the moment that for our matrix A the roots of the characteristic polynomial are distinct, $\lambda_i \neq \lambda_j$ if $i \neq j$.

Theorem. *If the characteristic polynomial of a complex $n \times n$-matrix A has n distinct roots $\lambda_1, ..., \lambda_n$ then the matrix is similar to the diagonal matrix* $\Lambda = \begin{bmatrix} \lambda_1 & 0 & ... \\ & ... & \\ ... & 0 & \lambda_n \end{bmatrix}$*: there exists an invertible complex $n \times n$-matrix C such that* $C^{-1}AC = \Lambda$.

Proof. Since $\det(\lambda_i I - A) = 0$, the system of linear equations $A\mathbf{z} = \lambda_i \mathbf{z}$ (with complex coefficients) has a non-trivial solution $\mathbf{z}_i \neq \mathbf{0}$. We claim that the complex eigenvectors $\mathbf{z}_1, ..., \mathbf{z}_n$ corresponding to the distinct eigenvalues $\lambda_1, ..., \lambda_n$ are linearly independent and therefore form a basis in $\mathbb{C}^n$. Then the matrix C whose columns are $\mathbf{z}_1, ..., \mathbf{z}_n$ is invertible, and we have $AC = C\Lambda$ and hence $C^{-1}AC = \Lambda$.

In order to justify the claim, let us assume that, say, $\mathbf{z}_1, ..., \mathbf{z}_k$ are linearly independent, but $\mathbf{z}_{k+1}$ is their linear combination. Consider the subspace $V = Span(\mathbf{z}_1, ...\mathbf{z}_k)$ and the linear transformation from V to itself defined by A: $A\mathbf{z}_i = \lambda_i \mathbf{z}_i$, $i = 1, ..., k$. The matrix of this linear transformation in the basis $\mathbf{z}_1, ..., \mathbf{z}_k$ is diagonal with the characteristic polynomial $(\lambda - \lambda_1)...(\lambda - \lambda_k)$. Since λ_{k+1} is not a root of this characteristic polynomial, our linear transformation from V to itself can not have an eigenvector with the eigenvalue λ_{k+1}. Thus $\mathbf{z}_{k+1}$ does not belong to V in contradiction with our assumption. This contradiction completes the proof of the diagonalization theorem.

Let us improve the result in the case when the matrix A is actually real. In this case the characteristic polynomial of A has real coefficients and therefore its roots are either real or come in complex conjugated pairs $\lambda, \bar{\lambda}$. For a real root λ_i the corresponding eigenvector can be taken real. For a complex conjugated pair, if a complex non-zero vector $\mathbf{z} = (z_1, ..., z_n)$ satisfies $A\mathbf{z} = \lambda\mathbf{z}$ then $A\bar{\mathbf{z}} = \bar{\lambda}\bar{\mathbf{z}}$ so

[3] $\mathbb{C}$ is the standard notation for the set of complex numbers.

that $\bar{\mathbf{z}} = (\bar{z}_1, ..., \bar{z}_n)$ is an eigenvector with the eigenvalue $\bar{\lambda}$. As we know from Section 1.5.3, the linear transformation A acts on the plane spanned by the real and imaginary parts of the vector $\mathbf{z}$ as the complex multiplication by $\lambda = a + bi$ and has the matrix $\begin{bmatrix} a & -b \\ b & a \end{bmatrix}$. We arrive therefore at the following conclusion.

Corollary. *Suppose that the characteristic polynomial of a linear transformation from $\mathbb{R}^n$ to itself has $n = r + 2s$ distinct roots: r real roots $\lambda_1, ..., \lambda_r$ and s pairs of complex conjugated roots $a_1 \pm b_1 i, ..., a_s \pm b_s i$. Then in a suitable coordinate system in $\mathbb{R}^n$ the linear transformation has the matrix*

$$\begin{bmatrix} \lambda_1 & & & & & & & \\ & \cdots & & & & & & \\ & & \lambda_r & & & & & \\ & & & a_1 & -b_1 & & & \\ & & & b_1 & a_1 & & & \\ & & & & & \cdots & & \\ & & & & & & a_s & -b_s \\ & & & & & & b_s & a_s \end{bmatrix}.$$

What happens if the characteristic polynomial of the complex matrix A has multiple roots?

Examples. (a) The $m \times m$-matrix

$$J_{\lambda_0,m} = \begin{bmatrix} \lambda_0 & 1 & 0 & \dots & 0 \\ 0 & \lambda_0 & 1 & 0 & \dots \\ & & \dots & & \\ 0 & \dots & 0 & \lambda_0 & 1 \\ 0 & \dots & & 0 & \lambda_0 \end{bmatrix}$$

has the characteristic polynomial $(\lambda - \lambda_0)^m$ with one root λ_0 of multiplicity m. It is called the **Jordan cell** of size m with the eigenvalue λ_0 (which can be a complex number). The linear transformation defined by the matrix $J_{\lambda_0,m}$ acts on the unit coordinate vectors in the following way:

$$\mathbf{e}_m \mapsto \lambda_0 \mathbf{e}_m + \mathbf{e}_{m-1}, \ \mathbf{e}_{m-1} \mapsto \lambda_0 \mathbf{e}_{m-1} + \mathbf{e}_{m-2}, \ ..., \ \mathbf{e}_2 \mapsto \lambda_0 \mathbf{e}_2 + \mathbf{e}_1, \ \mathbf{e}_1 \mapsto \lambda_0 \mathbf{e}_1.$$

In particular, $\mathbf{e}_1$ is an eigenvector with the eigenvalue λ_0. In fact any eigenvector of $J_{\lambda_0,m}$ is proportional to $\mathbf{e}_1$. Indeed, the matrix $J_{\lambda_0,m} - \lambda_0 I$ has rank $m - 1$ (since the last $m - 1$ columns of this matrix are equal to $\mathbf{e}_1, ..., \mathbf{e}_{m-1}$ and are linearly independent). Hence the system $J_{\lambda_0,m}\mathbf{z} = \lambda_0 \mathbf{z}$ of linear equations has one-dimensional solution space. We see that the Jordan cell does not have a basis of eigenvectors and is therefore not similar to any diagonal matrix.

(b) Let $B_{\lambda_0,\mathbf{m}}$ be a block-diagonal matrix of the form

$$B_{\lambda_0,\mathbf{m}} = \begin{bmatrix} J_{\lambda_0,m_1} & \mathbf{0} & \dots & \mathbf{0} \\ \mathbf{0} & J_{\lambda_0,m_2} & \mathbf{0} & \dots \\ & \dots & & \\ \mathbf{0} & \dots & \mathbf{0} & J_{\lambda_0,m_s} \end{bmatrix},$$

with Jordan cells of sizes $m_1 \geq m_2 \geq ... \geq m_s \geq 1$ on the role of the diagonal blocks. We assume here that all the Jordan cells have the same eigenvalue λ_0, and that the sizes add up to m. We use the symbol $\mathbf{m}$ to denote the whole string of sizes $(m_1, ..., m_s)$.

The characteristic polynomial of the matrix $B_{\lambda_0,\mathbf{m}}$ is the product $(\lambda - \lambda_0)^{m_1} \ldots (\lambda - \lambda_0)^{m_s} = (\lambda - \lambda_0)^m$ of the characteristic polynomials of the Jordan cells. It is easy to see that the eigenspace of $B_{\lambda_0,\mathbf{m}}$ has dimension s (one eigenline for each of the s Jordan cells). When $m_1 = \ldots = m_s = 1$ the matrix equals $\lambda_0 I_m$ and is diagonal. In any other case it has no basis of eigenvectors and is not similar to any diagonal matrix. In fact two matrices $B_{\lambda,\mathbf{m}}$ and $B_{\lambda',\mathbf{m}'}$ are similar if and only if $\lambda = \lambda'$ and $\mathbf{m} = \mathbf{m}'$, as it is stated in the following theorem.

Theorem. (Jordan normal forms.)
Suppose that the characteristic polynomial of a square matrix A has distinct complex roots $\lambda', \lambda'', \ldots$ of multiplicities $n', n'', \ldots$. Then A is similar to exactly one of the following block-diagonal matrices with the blocks of sizes $n', n'', \ldots$:

$$\begin{bmatrix} B_{\lambda',\mathbf{m}'} & \mathbf{0} & \ldots & \mathbf{0} \\ \mathbf{0} & B_{\lambda'',\mathbf{m}''} & \mathbf{0} & \ldots \\ \mathbf{0} & \mathbf{0} & \ldots & \end{bmatrix}.$$

Example. (c) According to the Jordan normal form theorem, a non-zero 4×4-matrix with the characteristic polynomial λ^4 is similar to exactly one of the following matrices:

$$\begin{bmatrix} 0&1&0&0\\0&0&1&0\\0&0&0&1\\0&0&0&0 \end{bmatrix}, \begin{bmatrix} 0&1&0&0\\0&0&1&0\\0&0&0&0\\0&0&0&0 \end{bmatrix}, \begin{bmatrix} 0&1&0&0\\0&0&0&0\\0&0&0&1\\0&0&0&0 \end{bmatrix}, \begin{bmatrix} 0&1&0&0\\0&0&0&0\\0&0&0&0\\0&0&0&0 \end{bmatrix}.$$

We stop here in our discussion of similarity classification of matrices and leave a proof of the Jordan theorem to a more systematic linear algebra course.

Exercises 3.6.1.
Are the following matrices diagonalizable?

$$(a) \begin{bmatrix} 9 & 22 & -6 \\ -1 & -4 & 1 \\ 8 & 16 & -5 \end{bmatrix}, \quad (b) \begin{bmatrix} 0 & 3 & 3 \\ -1 & -8 & 6 \\ 2 & -14 & -10 \end{bmatrix}.$$

Find Jordan normal forms of the following matrices

$$(c) \begin{bmatrix} 1&2&3&4\\0&1&2&3\\0&0&1&2\\0&0&0&1 \end{bmatrix}, \quad (d) \begin{bmatrix} 0&1&0&0\\0&0&1&0\\0&0&0&1\\1&0&0&0 \end{bmatrix}.$$

Diagonalize the following matrices by a real or complex similarity transformation:

$$(e) \begin{bmatrix} 1 & 2 & 0 \\ 0 & 2 & 0 \\ -2 & -2 & -1 \end{bmatrix}, \quad (f) \begin{bmatrix} 4 & 6 & 0 \\ -3 & -5 & 0 \\ -3 & -6 & 1 \end{bmatrix}, \quad (g) \begin{bmatrix} 7 & -12 & -2 \\ 3 & -4 & 0 \\ -2 & 0 & -2 \end{bmatrix},$$

$$(h) \begin{bmatrix} -2 & 8 & 6 \\ -4 & 10 & 6 \\ 4 & -8 & -4 \end{bmatrix}, \quad (i) \begin{bmatrix} 3 & 7 & -3 \\ -2 & -5 & 2 \\ -4 & -10 & 3 \end{bmatrix}, \quad (j) \begin{bmatrix} 1 & -1 & 2 \\ 3 & -3 & 6 \\ 2 & -2 & 4 \end{bmatrix}.$$

(k) List all Jordan normal forms with characteristic polynomials $(\lambda+1)^4$, $(\lambda^2-1)^2$, λ^5.
(l) Prove that a complex matrix A satisfying $A^{1999} = I$ is diagonalizable.
(m) Prove that transposed square matrices are similar to each other.

3.6.2. Linear ODE systems. Let

$$\begin{aligned} \dot{x}_1 &= a_{11}x_1 + ... + a_{1n}x_n \\ &\ldots \\ \dot{x}_2 &= a_{n1}x_1 + ... + a_{nn}x_n \end{aligned}$$

be a linear homogeneous system of ordinary differential equations with constant complex coefficients a_{ij}. It can be written in the matrix form as $\dot{\mathbf{x}} = A\mathbf{x}$.

Consider the infinite matrix series

$$e^{tA} := I + tA + \frac{t^2A^2}{2} + \frac{t^3A^3}{6} + ... + \frac{t^kA^k}{k!} + ...$$

If M is an upper bound for the absolute values of the entries of A, then the matrix entries of $t^kA^k/k!$ are bounded by $n^kt^kM^k/k!$. It is easy to deduce from this that the series converges (at least as fast as the series for $\exp(ntM)$).

Proposition. *The solution to the system* $\dot{\mathbf{x}} = A\mathbf{x}$ *with the initial condition* $\mathbf{x}(0)$ *is given by the formula* $\mathbf{x}(t) = e^{tA}\mathbf{x}(0)$.

Proof. Differentiating the series $\sum_0^\infty t^kA^k/k!$ we find

$$\frac{d}{dt}e^{tA} = \sum_{k=1}^{\infty}\frac{t^{k-1}A^k}{(k-1)!} = \sum_{k=0}^{\infty}\frac{t^kA^{k+1}}{k!} = Ae^{tA}$$

and hence $\frac{d}{dt}e^{tA}\mathbf{x}(0) = A(e^{tA}\mathbf{x}(0))$. Thus $\mathbf{x}(t)$ satisfies the ODE system. At $t = 0$ we have $e^{0A}\mathbf{x}(0) = I\mathbf{x}(0) = \mathbf{x}(0)$ and therefore the initial condition is also satisfied.

The proposition reduces the problem of solving the ODE system $\dot{\mathbf{x}} = A\mathbf{x}$ to computation of the exponential function e^{tA} of a matrix. Notice that if $A = CBC^{-1}$ then

$$A^k = CBC^{-1}CBC^{-1}CBC^{-1}... = CBBB...C^{-1} = CB^kC^{-1}$$

and therefore $\exp(tA) = C^{-1}\exp(tB)C$. This observation reduces computation of e^{tA} to that of e^{tB} where the Jordan normal form of A can be taken on the role of B.

Examples. (a) Let Λ be a diagonal matrix with the diagonal entries $\lambda_1, ..., \lambda_n$. Then Λ^k is a diagonal matrix with the diagonal entries $\lambda_1^k, ..., \lambda_n^k$ and hence

$$e^{t\Lambda} = I + t\Lambda + \frac{t^2}{2}\Lambda^2 + ... = \begin{bmatrix} e^{\lambda_1 t} & 0 & ... \\ & ... & \\ ... & 0 & e^{\lambda_n t} \end{bmatrix}.$$

(b) Let N be a Jordan cell of size m with zero eigenvalue. We have (for $m = 4$)

$$N = \begin{bmatrix} 0&1&0&0\\0&0&1&0\\0&0&0&1\\0&0&0&0 \end{bmatrix},\ N^2 = \begin{bmatrix} 0&0&1&0\\0&0&0&1\\0&0&0&0\\0&0&0&0 \end{bmatrix},\ N^3 = \begin{bmatrix} 0&0&0&1\\0&0&0&0\\0&0&0&0\\0&0&0&0 \end{bmatrix},\ N^4 = \mathbf{0}.$$

Following the pattern we find

$$e^{tN} = I + tN + \frac{t^2}{2}N^2 + ... + \frac{t^{m-1}}{(m-1)!}N^{m-1} = \begin{bmatrix} 1 & t & \frac{t^2}{2} & ... & \frac{t^{m-1}}{(m-1)!} \\ 0 & 1 & t & ... & \frac{t^{m-2}}{(m-2)!} \\ 0 & 0 & 1 & t & ... \\ & & ... & & \\ & ... & & 0 & 1 \end{bmatrix}.$$

(c) Let $\lambda I + N$ be the Jordan cell of size m with the eigenvalue λ. Then $e^{t(\lambda I+N)} = e^{t\lambda I}e^{tN} = e^{\lambda t}e^{tN}$. Here we use the multiplicative property of the matrix exponential function: [4]

$$e^{A+B} = e^A e^B \textit{ provided that } A \textit{ and } B \textit{ commute: } AB = BA.$$

(d) Let $A = \begin{bmatrix} B & \mathbf{0} \\ \mathbf{0} & D \end{bmatrix}$ be a block-diagonal square matrix. Then $A^k = \begin{bmatrix} B^k & \mathbf{0} \\ \mathbf{0} & D^k \end{bmatrix}$ and respectively $e^{tA} = \begin{bmatrix} e^{tB} & \mathbf{0} \\ \mathbf{0} & e^{tD} \end{bmatrix}$. Together with the Examples (b) and (c) this shows how to compute the exponential function e^{tJ} for any Jordan normal form J: each Jordan cell has the form $\lambda I+N$ and should be replaced by $e^{\lambda t}e^{tN}$. Since any square matrix A can be reduced to one of the Jordan normal matrices J by similarity transformations, we conclude that $e^{tA} = Ce^{tJ}C^{-1}$ with suitable invertible C.

Applying the Example (a) to ODE systems $\dot{\mathbf{x}} = A\mathbf{x}$ we arrive at the following conclusion. Suppose that the characteristic polynomial of A has n distinct roots $\lambda_1, ..., \lambda_n$. Let C be the matrix whose columns are the corresponding n complex eigenvectors $\mathbf{z}_1, ..., \mathbf{z}_n$. Then the solution with the initial condition $\mathbf{x}(0)$ is given by the formula

$$\mathbf{x}(t) = C \begin{bmatrix} e^{\lambda_1 t} & 0 & \dots \\ & \dots & \\ \dots & 0 & e^{\lambda_n t} \end{bmatrix} C^{-1}\mathbf{x}(0).$$

Notice that $C^{-1}\mathbf{x}(0)$ here (as well as $\mathbf{x}(0)$) is a column $\mathbf{c} = (c_1, ..., c_n)^t$ of arbitrary constants, and the columns of $Ce^{t\Lambda}$ are $e^{\lambda_i t}\mathbf{z}_i$. We conclude that the general solution formula reads

$$\mathbf{x}(t) = c_1 e^{\lambda_1 t}\mathbf{z}_1 + ... + c_n e^{\lambda_n t}\mathbf{z}_n.$$

This formula involves eigenvalues λ_i of A, the corresponding eigenvectors $\mathbf{z}_i$ and the arbitrary complex constants c_i. The values of $c_1, ..., c_n$ corresponding to a given initial condition $\mathbf{x}(0)$ can be found by representing $\mathbf{x}(0)$ as a linear combination $c_1\mathbf{z}_1 + ... + c_n\mathbf{z}_n$ of the eigenvectors.

Example. (e) The ODE system

$$\begin{array}{rcl} \dot{x}_1 & = & 2x_1 + x_2 \\ \dot{x}_2 & = & x_1 + 3x_2 - x_3 \\ \dot{x}_3 & = & 2x_2 + 3x_3 - x_1 \end{array} \quad \text{has } A = \begin{bmatrix} 2 & 1 & 0 \\ 1 & 3 & -1 \\ -1 & 2 & 3 \end{bmatrix}.$$

The characteristic polynomial $\lambda^3 - 8\lambda^2 + 22\lambda - 20$ has a real root $\lambda_1 = 2$. It factors as $(\lambda-2)(\lambda^2-6\lambda+10)$ and thus has two complex roots $\lambda_2 = 3+i$, $\lambda_3 = 3-i$. The 1-st eigenvector $\mathbf{z}_1 = (1, 0, 1)$ is found from the system $2x_1 + x_2 = 2x_1$, $x_1 + 3x_2 - x_3 = 2x_2$, $-x_1 + 2x_2 + 3x_3 = 2x_3$. The 2-nd eigenvector $\mathbf{z}_2 = (1, 1+i, 2-i)$ is found from the system $2x_1+x_2 = (3+i)x_1$, $x_1+3x_2-x_3 = (3+i)x_2$, $-x_1+2x_2+3x_3 = (3+i)x_3$. The complex conjugate $\mathbf{z}_3 = (1, 1-i, 2+i)$ to $\mathbf{z}_2$ can be taken on the role of the

[4]This is just a property of the exponential series and can be proved in exactly the same way as the multiplicativity of the complex exponential function $\exp(z+w) = (\exp z)(\exp w)$ in Section 1.4.2. As one can see from that proof, the commutativity $AB = BA$ is essential. When $AB \neq BA$, the property usually fails, and we recommend the reader to find an example where $e^{A+B} \neq e^A e^B$.

3-rd eigenvector. Thus the general complex solution is a linear combination of the solutions

$$e^{2t}\begin{bmatrix}1\\0\\1\end{bmatrix},\ e^{(3+i)t}\begin{bmatrix}1\\1+i\\2-i\end{bmatrix},\ e^{(3-i)t}\begin{bmatrix}1\\1-i\\2+i\end{bmatrix}$$

with arbitrary complex coefficients. Real solutions can be extracted from the complex ones by taking their real and imaginary parts:

$$e^{3t}\begin{bmatrix}\cos t\\ \cos t-\sin t\\ 2\cos t+\sin t\end{bmatrix},\ e^{3t}\begin{bmatrix}\sin t\\ \cos t+\sin t\\ 2\sin t-\cos t\end{bmatrix}.$$

Thus the general real solution is described by the formulas

$$\begin{array}{lclclcl} x_1(t) & = & c_1e^t & + & c_2e^{3t}\cos t & + & c_3e^{3t}\sin t \\ x_2(t) & = & & & c_2e^{3t}(\cos t-\sin t) & + & c_3e^{3t}(2\cos t+\sin t) \\ x_3(t) & = & c_1e^t & + & c_2e^{3t}(\cos t+\sin t) & + & c_3e^{3t}(2\sin t-\cos t) \end{array},$$

where c_1, c_2, c_3 are arbitrary *real* constants. At $t = 0$ we have

$$x_1(0) = c_1 + c_2,\ x_2(0) = c_2 + 2c_3,\ x_3(0) = c_1 + c_2 - c_3.$$

Given the initial values $(x_1(0), x_2(0), x_3(0))$ the corresponding constants c_1, c_2, c_3 can be found from this system of linear algebraic equations.

In general, even if the characteristic polynomial of A has multiple roots, the Examples (a) – (d) show that solutions to the ODE system $\dot{\mathbf{x}} = A\mathbf{x}$ are expressible as linear combinations of the functions $t^ke^{\lambda t}$, $t^ke^{at}\cos bt$, $t^ke^{at}\sin bt$ where λ are real eigenvalues, $a \pm ib$ are complex eigenvalues, and $k = 0, 1, 2, \ldots$ is to be smaller than the multiplicity of the corresponding eigenvalue. This observation suggests to approach the ODE systems with multiple eigenvalues in the following way avoiding explicit similarity transformation to the Jordan normal form: look for the general solution in the form of linear combinations of these functions with arbitrary coefficients by substituting them into the equations and find the relations between the arbitrary constants from the resulting system of linear algebraic equations.

Example. (f) The ODE system

$$\begin{array}{lcl} \dot{x}_1 & = & 2x_1 + x_2 + x_3 \\ \dot{x}_2 & = & -3x_1 - 2x_2 - 3x_3 \\ \dot{x}_3 & = & 2x_1 + 2x_2 + 3x_3 \end{array},\quad \text{has } A = \begin{bmatrix} 2 & 1 & 1 \\ -3 & -2 & -3 \\ 2 & 2 & 3 \end{bmatrix}$$

with the characteristic polynomial $\lambda^3 - 3\lambda^2 + 3\lambda - 1 = (\lambda - 1)^3$. Thus we can look for solutions in the form
$x_1 = e^t(a_1 + b_1t + c_1t^2)$, $x_2 = e^t(a_2 + b_2t + c_2t^2)$, $x_3 = e^t(a_3 + b_3t + c_3t^2)$.
Substituting into the ODE system (and omitting the factors e^t), we get

$$(a_1 + b_1) + (b_1 + 2c_1)t + c_1t^2 =$$

$$= (2a_1 + a_2 + a_3) + (2b_1 + b_2 + b_3)t + (2c_1 + c_2 + c_3)t^2,$$

$$(a_2 + b_2) + (b_2 + 2c_2)t + c_2t^2 =$$

$$= -(3a_1 + 2a_2 + 3a_3) - (3b_1 + 2b_2 + 3b_3)t - (3c_1 + 2c_2 + 3c_3)t^2,$$

$$(a_3 + b_3) + (b_3 + 2c_3)t + c_3t^2 =$$

$$= (2a_1 + 2a_2 + 3a_3) + (2b_1 + 2b_2 + 3b_3)t + (2c_1 + 2c_2 + 3c_3)t^2.$$

Introducing the notation $A = \sum a_i, B = \sum b_i,\ C = \sum c_i,\ P = A + Bt + Ct^2$, we rewrite the system of 9 linear equations in 9 unknowns in the form

$$b_1 + 2c_1 t = P,\ b_2 + 2c_2 t = -3P,\ b_3 + 2c_3 t = 2P.$$

This yields $b_1 = A, b_2 = -3A, b_3 = 2A$ and hence $c_1 = c_2 = c_3 = 0$ since $B = A - 3A + 2A = 0$. The general solution to the ODE system is therefore found:

$$\begin{bmatrix} x_1 \\ x_2 \\ x_3 \end{bmatrix} = \begin{bmatrix} e^t(a_1 + (a_1 + a_2 + a_3)t) \\ e^t(a_2 - 3(a_1 + a_2 + a_3)t) \\ e^t(a_3 + 2(a_1 + a_2 + a_3)t) \end{bmatrix},$$

where a_1, a_2, a_3 are arbitrary constants. [5]

Exercises 3.6.2.

Solve the following ODE systems. Find the solution satisfying the initial condition $x_1(0) = 1,\ x_2(0) = 0,\ x_3(0) = 0$.

$$(a)\ \begin{array}{l} \dot{x}_1 = 3x_1 - x_2 + x_3 \\ \dot{x}_2 = x_1 + x_2 + x_3 \\ \dot{x}_3 = 4x_1 - x_2 + 4x_3 \\ (\lambda_1 = 1) \end{array} \quad (b)\ \begin{array}{l} \dot{x}_1 = -3x_1 + 4x_2 - 2x_3 \\ \dot{x}_2 = x_1 + x_3 \\ \dot{x}3 = 6x_1 - 6x_2 + 5x_3 \\ (\lambda_1 = 1) \end{array} \quad (c)\ \begin{array}{l} \dot{x}_1 = x_1 - x_2 - x_3 \\ \dot{x}_2 = x_1 + x_2 \\ \dot{x}_3 = 3x_1 + x_3 \\ (\lambda_1 = 1) \end{array}$$

$$(d)\ \begin{array}{l} \dot{x}_1 = 4x_1 - x_2 - x_3 \\ \dot{x}_2 = x_1 + 2x_2 - x_3 \\ \dot{x}_3 = x_1 - x_2 + 2x_3 \\ (\lambda_1 = 2) \end{array} \quad (e)\ \begin{array}{l} \dot{x}_1 = -x_1 + x_2 - 2x_3 \\ \dot{x}_2 = 4x_1 + x_2 \\ \dot{x}_3 = 2x_1 + x_2 - x_3 \\ (\lambda_1 = 1) \end{array}$$

$$(f)\ \begin{array}{l} \dot{x}_1 = 4x_1 - x_2 \\ \dot{x}_2 = 3x_1 + x_2 - x_3 \\ \dot{x}_3 = x_1 + x_3 \\ (\lambda_1 = 2) \end{array} \quad (g)\ \begin{array}{l} \dot{x}_1 = 2x_1 - x_2 - x_3 \\ \dot{x}_2 = 2x_1 - x_2 - 2x_3 \\ \dot{x}_3 = -x_1 + x_2 + 2x_3 \\ (\lambda_1 = 1) \end{array}$$

(h) Sketch $3D$ phase portraits of the linear ODE systems $\begin{array}{l} \dot{x}_1 = \lambda_1 x_1 \\ \dot{x}_2 = \lambda_2 x_2 \\ \dot{x}_3 = \lambda_3 x_3 \end{array}$ with

$$\lambda_1 > \lambda_2 > \lambda_3 > 0,\ \ \lambda_1 > \lambda_2 > 0 > \lambda_3,\ \ \lambda_1 > 0 > \lambda_2 > \lambda_3,\ \ 0 > \lambda_1 > \lambda_2 > \lambda_3.$$

Is the equilibrium $\mathbf{x} = \mathbf{0}$ of these systems asymptotically stable?

[5] In particular, since t^2 does not occur in the formulas, we can conclude that the Jordan form of our matrix has only Jordan cells of size 1 or 2, and hence — one cell of size 1 and one — of size 2 since the total size of the matrix is 3: $\begin{bmatrix} 1 & 1 & 0 \\ 0 & 1 & 0 \\ 0 & 0 & 1 \end{bmatrix}$.

3.6.3. Higher order linear ODEs. A linear homogeneous constant coefficient n-th order ODE

$$\frac{d^n}{dt^n}x + a_1\frac{d^{n-1}}{dt^{n-1}}x + ... + a_{n-1}\frac{d}{dt}x + a_n x = 0$$

can be rewritten as a system $\dot{\mathbf{x}} = A\mathbf{x}$ of n first order equations with the matrix

$$A = \begin{bmatrix} 0 & 1 & 0 & ... & 0 \\ 0 & 0 & 1 & 0 & ... \\ & & ... & & \\ 0 & ... & 0 & 0 & 1 \\ -a_n & -a_{n-1} & ... & -a_2 & -a_1 \end{bmatrix},$$

by introducing the notations $x_1 = x, x_2 = \dot{x}, x_3 = \ddot{x}, ..., x_n = d^{n-1}x/dt^{n-1}$. Then our theory of linear ODE system applies. There are however some simplifications which are due to the special form of the matrix A. First, computing the characteristic polynomial of A we find

$$\det(\lambda I - A) = \lambda^n + a_1\lambda^{n-1} + ... + a_{n-1}\lambda + a_n.$$

Thus the polynomial can be easily read off the high order differential equation. Next, let $\lambda_1, ..., \lambda_r$ be the roots of the characteristic polynomial, and $m_1, ..., m_r$ — their multiplicities ($m_1 + ... + m_r = n$). Then it is clear that the solutions $x(t) = x_1(t)$ must have the form

$$e^{\lambda_1 t}P_1(t) + ... + e^{\lambda_r t}P_r(t),$$

where $P_i = a_0 + a_1 t + ... + a_{m_i - 1}t^{m_i - 1}$ is a polynomial of degree $< m_i$. The total number of arbitrary coefficients in these polynomials equals $m_1 + ... + m_r = n$. On the other hand, the general solution to the n-th order ODE must depend on n arbitrary initial values $(x(0), \dot{x}(0), ..., x^{(n-1)}(0))$ due to the Existence and Uniqueness Theorem. We conclude that *all the n functions*

$$e^{\lambda_1 t},\ e^{\lambda_1 t}t,\ ...,\ e^{\lambda_1 t}t^{m_1-1},\ ...,\ e^{\lambda_r t},\ e^{\lambda_r t}t,\ ...,\ e^{\lambda_r t}t^{m_r-1}$$

must satisfy our differential equation, and any (complex) solution is uniquely written as a linear combination of these functions with suitable (complex) coefficients. (In other words — these functions form a basis of complex solutions to the differential equation.)

Example. The differential equation $x^{(xii)} - 3x^{(viii)} + 3x^{(iv)} - x = 0$ has the characteristic polynomial $\lambda^{12} - 3\lambda^8 + 3\lambda^4 - 1 = (\lambda - 1)^3(\lambda + 1)^3(\lambda - i)^3(\lambda + i)^3$. The following 12 functions form therefore a basis of complex solutions:

$$e^t,\ te^t,\ t^2e^t,\ e^{-t},\ te^{-t},\ t^2e^{-t},\ e^{it},\ te^{it},\ t^2e^{it},\ e^{-it},\ te^{-it},\ t^2e^{-it}.$$

Of course, a basis of real solutions is obtained by taking real and imaginary parts of complex solutions:

$$e^t,\ te^t,\ t^2e^t,\ e^{-t},\ te^{-t},\ t^2e^{-t},\ \cos t,\ \sin t,\ t\cos t,\ t\sin t,\ t^2\cos t,\ t^2\sin t.$$

Remark. The fact that the functions $e^{\lambda_i t}t^k$, $k < m_i$, $i = 1, ..., r$, form a basis of solutions to the differential equation with the characteristic polynomial $(\lambda - \lambda_1)^{m_1}...(\lambda - \lambda_r)^{m_r}$ is not hard to check directly, without a reference to linear algebra and Existence and Uniqueness Theorem. However, it is useful to understand how this property of the equation is related to the Jordan structure of the corresponding matrix A. In fact the Jordan normal form of the matrix A consists of exactly r

Jordan cells — one cell of size m_i for each eigenvalue λ_i. This simplification can be explained as follows. For any λ the matrix $\lambda I - A$ has rank $n-1$ *at least* (due to the presence of the $(n-1)\times(n-1)$ identity submatrix in the right upper corner of A). This guarantees that the eigenspaces of A have dimension 1 *at most* and hence A cannot have more than one Jordan cell corresponding to the same root of the characteristic polynomial. Using this property, the reader can check now that the formulation of the Jordan Theorem in terms of differential equations given in Section 3.1 is indeed equivalent to the matrix formulation given in Section 3.6.1.

Exercises 3.6.3.

Solve the following ODEs. Find the solution satisfying the initial condition $x(0)=1, \dot{x}(0)=0, ..., x^{(n-1)}(0)=0$.

(a) $x^{(iii)} - 8x = 0$

(b) $x^{(iv)} + 4x = 0$

(c) $x^{(vi)} + 64x = 0$

(d) $x^{(v)} - 10x^{(iii)} + 9x = 0$

(e) $x^{(iii)} - 3x^{(i)} - 2x = 0$

(f) $x^{(v)} - 6x^{(iv)} + x^{(iii)} = 0$

(g) $x^{(v)} + 8x^{(iii)} + 16x^{(i)} = 0$

(h) $x^{(iv)} + 4x^{(ii)} + 3x = 0$

(i) Rewrite the ODE system

$$\begin{aligned} \ddot{x}_1 + 4\dot{x}_1 - 2x_1 - 2\dot{x}_2 - x_2 = 0 \\ \ddot{x}_1 - 4\dot{x}_1 - \ddot{x}_2 + 2\dot{x}_2 + 2x_2 = 0 \end{aligned}$$

of two 2-nd order equations in the form of a linear ODE system $\dot{\mathbf{x}} = A\mathbf{x}$ of four 1-st order equations and solve it.

3.7. Vector spaces

In this course, we have encountered linear combinations and inner products of different objects — geometrical vectors, columns of numbers, trigonometric functions, polynomials, solutions of ODEs — and have noted certain similarity in their properties. To understand this similarity, we will treat these different objects as particular examples of *vectors* in abstract *vector spaces* which are defined axiomaticly by way of listing their formal properties but without any reference to the actual nature of the objects.

3.7.1. Axioms and examples. *Definition.* A vector space is a set V (whose elements are called vectors) provided with operations of addition and multiplication by scalars which satisfy the following conditions:

(i) the sum of two vectors $\mathbf{u}$ and $\mathbf{v}$ is a vector (denoted $\mathbf{u}+\mathbf{v}$); the result of multiplication of a vector $\mathbf{v}$ by a scalar λ is a vector (denoted $\lambda\mathbf{v}$);

(ii) addition of vectors is commutative and associative: for any $\mathbf{u}, \mathbf{v}, \mathbf{w}$ from V we have

$$\mathbf{u}+\mathbf{v}=\mathbf{v}+\mathbf{u}, \quad (\mathbf{u}+\mathbf{v})+\mathbf{w}=\mathbf{u}+(\mathbf{v}+\mathbf{w});$$

(iii) there exists the zero vector (denoted $\mathbf{0}$) such that

$$\mathbf{v}+\mathbf{0}=\mathbf{0}+\mathbf{v}=\mathbf{v} \quad \text{for any } \mathbf{v} \text{ from } V;$$

(iv) for any vector $\mathbf{u}$ there exists the opposite vector (denoted $-\mathbf{u}$) such that

$$-\mathbf{u}+\mathbf{u}=\mathbf{0};$$

(v) the multiplication by scalars is distributive: for any vectors $\mathbf{u}, \mathbf{v}$ and any scalars λ, μ we have

$$(\lambda+\mu)(\mathbf{u}+\mathbf{v})=\lambda\mathbf{u}+\lambda\mathbf{v}+\mu\mathbf{u}+\mu\mathbf{v};$$

(vi) the multiplication is associative in the following sense: for any vector $\mathbf{v}$ and any scalars λ, μ we have:

$$(\lambda\mu)\mathbf{u}=\lambda(\mu\mathbf{u});$$

(vii) the multiplication by the scalars 0 and 1 act on any vector $\mathbf{u}$ as

$$0\mathbf{u}=\mathbf{0}, \quad 1\mathbf{u}=\mathbf{u}.$$

We have to add to this definition the following comment about the use of the word *scalars*. We can take one of the sets $\mathbb{R}$ or $\mathbb{C}$ of real or complex numbers on the role of scalars and obtain the definition of a real vector space or a complex vector space. In fact *any* system $\mathbb{K}$ of "numbers" closed with respect to addition and multiplication will do, *provided that the multiplication of numbers is commutative and that the division by all non-zero numbers is defined.* [6] We arrive in this way to the notion of $\mathbb{K}$-vector spaces. For instance, the set $\mathbb{Z}_2=\{0,1\}$ of remainders modulo 2 with the usual arithmetics of remainders ($0+0=0=1+1, 0+1=1=1+0, 0\cdot 0=1\cdot 0=0\cdot 1=0, 1\cdot 1=1$) can be taken on the role of scalars which gives rise to the definition of $\mathbb{Z}_2$-vector spaces useful in computer science and logic.

[6] For example, the set of all integers, or of all polynomials is not allowed on the role of scalars since the division is not always defined, but all rational numbers or all rational functions $P(x)/Q(x)$ on the role of $\mathbb{K}$ are OK.

Before presenting examples of vector spaces we would like to emphasize that the axioms (i) – (vii) describe properties of *operations* with vectors rather than properties of vectors themselves. In order to find different examples of vector spaces we therefore should figure out which operations are good candidates on the role of addition and multiplication by scalars. It turns out that in almost all useful examples the operations are *pointwise operations of addition and scalar multiplication of functions.*

Examples. (a) Let S be *any* set and V be the set $\mathbb{K}[S]$ of *all* functions on S with values in $\mathbb{K}$ provided with pointwise operations:

$$(f+g)(s) = f(s)+g(s), \ (\lambda f)(s) = \lambda(f(s)).$$

Then V is a $\mathbb{K}$-vector space.

(b) Let S be the set of n elements $1, 2, ..., n$. Then $\mathbb{R}[S] = \mathbb{R}^n$ and $\mathbb{C}[S] = \mathbb{C}^n$: each function $\mathbf{x}$ is specified by a column $(x(1), ..., x(n))^t$ of its values, and the operations are pointwise.

(c) Let S be the set of all ordered pairs (i, j) where $i = 1, ..., m$, $j = 1, ..., n$. Then the vector spaces $\mathbb{R}[S]$ and $\mathbb{C}[S]$ are respectively spaces of real and complex $m \times n$-matrices $[a(i, j)]$ with the usual operations of addition of matrices and their multiplication by scalars.

(d) Let V be a $\mathbb{K}$-vector space. Since elements of V can be added and multiplied by scalars, vector-valued functions from any set S to V can be added and multiplied pointwise. The set of all such functions is a $\mathbb{K}$-vector space too.

(e) A subset W in a vector space V is called a linear subspace if all linear combinations $\lambda\mathbf{u} + \mu\mathbf{v}$ of vectors from W are also in W. A subspace of a vector space satisfies the axioms of a vector space on its own (since the operations are the same as in V). For instance, all upper-triangular $n \times n$-matrices (lower-triangular, block-triangular, block-diagonal, diagonal matrices — the reader can continue this line of examples) form a subspace in the space of matrices and provide examples of vector spaces. The sets of all polynomials (say, in one variable), of all trigonometric polynomials, of all continuous (differentiable, 5 times continuously differentiable, infinitely differentiable) functions form a subspace in the space $\mathbb{R}[\mathbb{R}]$ of all real functions on the number line and therefore provide examples of vector spaces. Linear forms or quadratic forms in $\mathbb{R}^n$ form subspaces in the space $\mathbb{R}[\mathbb{R}^n]$ and therefore — vector spaces on their own.

(f) It is hard [7] to find examples of vector spaces which would not have the form of a subspace in the space of vector-valued functions on a set with pointwise operations of addition and scalar multiplication. It raises the suspicion that the axiomatic definition of vector spaces which obscures the actual nature of vectors as functions is useless. Here is an example of a construction of new vector spaces from old ones where it would be awkward to interpret the vectors as functions. In the theory of Fourier series we considered the space of $2L$-periodic piece-wise differentiable functions. However we ignored the values of the functions at discontinuity points as irrelevant. This means that we actually identified those functions on $\mathbb{R}$ which differ by their values at finitely many points. Thus, in the space V of all $2L$-periodic piece-wise differentiable functions on $\mathbb{R}$ (which is a subspace in $\mathbb{R}[\mathbb{R}]$) we consider the subspace W of all functions *non-zero* only at finitely many points on each period. A vector in the space of our interest corresponds to an *affine subspace*

[7]formally speaking — impossible

in V parallel to W. The set of all affine subspaces in a vector space V parallel to a given subspace W is a vector space V/W called the **quotient space** of V modulo W. Another example of this kind: let V be the space of all polynomials with real coefficients in one variable i, and W be the subspace of such polynomials divisible by i^2+1. Then it is natural to identify the quotient space V/W with the set $\mathbb{C}$ of complex numbers (which is indeed a 2-dimensional real vector space).

(g) There was however an example in this course where the vector operations did not originate from pointwise operations with functions: addition of geometrical vectors on the plane was introduced via composition of directed segments and *not* as pointwise addition of functions.

Definition. A vector space is called **finite-dimensional** if it can be spanned by finitely many vectors.

Let V be a finite-dimensional $\mathbb{K}$-vector space spanned by $\mathbf{v}_1, ..., \mathbf{v}_n$. We may assume that these vectors are linearly independent (otherwise we can throw away those of the vectors which are linear combinations of previous ones, and the remaining vectors will be linearly independent and will still span the whole space V). Then any vector $\mathbf{v}$ from V can be uniquely written as a linear combination $\mathbf{v} = x_1\mathbf{v}_1 + ... + x_n\mathbf{v}_n$. This identifies V with the space $\mathbb{K}^n$ of all functions from $\{1, ..., n\}$ to $\mathbb{K}$. [8] Then it is easy to prove (in the same way as we did it in Section 3.3 for $\mathbb{R}^n$) that any basis in V has exactly n elements (one announces this n to be the **dimension** of V), that any non-trivial subspace in V has finite dimension k, $0 < k < n$, etc. Thus, any geometrical fact about $\mathbb{R}^n$ remains true for any n-dimensional real vector space (and up to certain extent — for any n-dimensional $\mathbb{K}$-vector space).

The next section presents an example where geometrical intuition is applied to vectors in $\mathbb{Z}_2^n$.

Exercises 3.7.1.

(a) Deduce from the axioms (iv),(v),(vii) that $-\mathbf{u} = (-1)\mathbf{u}$.

(b) Verify the axioms (*i*) – (*vii*) for the space $\mathbb{C}[S]$ of all complex-valued functions on a set S.

(c) Let N be the null-space of a linear operator from $\mathbb{R}^n$ to $\mathbb{R}^m$. Identify the quotient space $\mathbb{R}^n/N$ with the column space of A in $\mathbb{R}^m$.

(d) Compute the dimension of the space of upper-triangular $n \times n$-matrices, of the space of linear forms in $\mathbb{R}^n$, of quadratic forms in $\mathbb{R}^n$, of bilinear forms in $\mathbb{R}^n$.

(e) Show that polynomials of degree n do not form a subspace in $\mathbb{R}[\mathbb{R}]$, while polynomials of degree $\leq n$ do. Compute its dimension.

(f) Prove that an infinite-dimensional vector space contains an infinite linearly independent set of vectors. Show that $\mathbb{K}[S]$ is finite-dimensional if and only if S is finite.

(g) Which of the examples of vector spaces given in the text are finite-dimensional?

3.7.2. Error-correcting codes. A piece of data is encoded by a string $(x_1, ..., x_n)$ of binary digits 0 or 1 and is wired from a transmitter to receiver. There is a small probability that the string is received with an error in one digit — 1 instead of 0 or 0 instead of 1, but simultaneous error in two bits is extremely unlikely. Thus, encoding the data by the $(n+1)$-string $(x_0, ..., x_n)$ where x_0 equals the parity of $x_1 + ... + x_n$ we give the addressee an opportunity to check whether the error took

[8] Do not overestimate this fact: the identification depends on the choice of the basis $\mathbf{v}_1, ..., \mathbf{v}_n$ and very often a "natural" choice is not unique. For instance, which basis would you choose in the space of polynomials of degree $\leq n$: $(1, t, t^2, ..., t^n)$, $(1, t-1, (t-1)^2, ..., (t-1)^n)$ $(1, t, t^2/2, ..., t^n/n!)$, or a basis $(L_0, ..., L_n)$ of Lagrange interpolation polynomials?

place: a string with $x_0 + ... + x_n = 1$ (in $\mathbb{Z}_2$) is erroneous, but if a string with $x_0 + ... + x_n = 0$ has arrived, an error is very unlikely.

How to improve this *parity check* (possibly — by sending a longer string) so that not only an error can be detected but also the bit where the error has occurred can be determined (and the error therefore — corrected) assuming that there is only one such a bit?

Let the string $\mathbf{x}$ (we consider it as a vector in $\mathbb{Z}_2^n$) represents the data, and the additional string we send be a vector $\mathbf{y}$ in $\mathbb{Z}_2^m$. The vector $\mathbf{x}$ can be arbitrary, but the vector $\mathbf{y}$ should depend on $\mathbf{x}$ in a certain way known to the addressee and thus provide the opportunity to detect an error by checking if the relationship between $\mathbf{x}$ and $\mathbf{y}$ is broken. Let us try a linear code, where $\mathbf{y} = A\mathbf{x}$ is obtained from $\mathbf{x}$ by a linear operator. Thus error-free strings $(\mathbf{x}, \mathbf{y})$ must have the form $(\mathbf{x}, A\mathbf{x})$. All such strings form the graph of the linear operator A from $\mathbb{Z}_2^n$ to $\mathbb{Z}_2^m$. The graph is a subspace V of dimension n in $\mathbb{Z}_2^{n+m}$.

What do we want from A? If an error occurs, the received vector $\mathbf{z} = (\mathbf{x}, \mathbf{y})$ differs by one bit from some vector in V and thus has the form $\mathbf{e}_j + (\mathbf{x}, A\mathbf{x})$ where $\mathbf{e}_j$ is one of the $m + n$ unit coordinate vectors [9] The erroneous vectors $\mathbf{z}$ form the affine subspaces passing through one of the points $\mathbf{e}_1, ..., \mathbf{e}_{m+n}$ and parallel to V. If two such affine subspaces meet at a point $\mathbf{z}$ (and hence coincide), it will be impossible to figure out which bit in $\mathbf{z}$ is erroneous. Thus what we need from A is that *all affine subspaces* $\mathbf{e}_1 + V, ..., \mathbf{e}_{n+m} + V$ *in* $\mathbb{Z}_2^{n+m}$ *parallel to the graph of* A *and passing through* $\mathbf{e}_1, ..., \mathbf{e}_{n+m}$ *are distinct (and differ from the graph* V *itself which corresponds to error-free strings).*

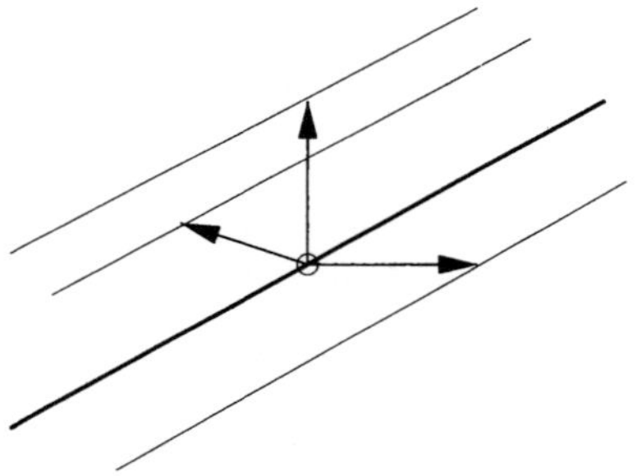

Affine subspaces parallel to the graph and passing through ***0, e₁ , e₂ , e₃,...*** *are to be distinct*

Notice that the graph is given by the equation $A\mathbf{x} = \mathbf{y}$ and is thus the null-space of the linear operator from $\mathbb{Z}_2^{n+m}$ to $\mathbb{Z}_2^m$ with the $m \times (n+m)$-matrix $[A|I_m]$. The parallel affine subspaces passing through $\mathbf{0}, \mathbf{e}_1, ..., \mathbf{e}_{m+n}$ are distinct if and only if the $m + n$ columns of the matrix $[A|I]$ are distinct and non-zero. This would be easy to accomplish by a suitable choice of the matrix A if the matrix were real. However our matrix consists of 0's and 1's. The total number of non-zero vectors in $\mathbb{Z}_2^m$ is finite and equal to $2^m - 1$, and m of them are columns of the identity matrix I_m. Thus we have totally $2^m - 1 - m$ different choices for columns of A,

[9] Ironically, the addressee would not know which part — $\mathbf{x}$ or $\mathbf{y}$ — contains the error.

and our problem has a solution only if $n \leq 2^m - 1 - m$. Since the relative length of the transmitted signal better be maximized, we arrive at the following description of economical error-correcting codes.

Pick $m > 2$ and put $n = 2^m - m - 1$ (we have $n = 4$ for $m = 3$, $n = 9$ for $m = 4$, $n = 26$ for $m = 5$, etc.) Take A to be any $m \times n$-matrix whose columns represent all m-columns of binary digits except $\mathbf{0}, \mathbf{e}_1, ..., \mathbf{e}_m$. For instance, if $m = 3$ take

$$A = \begin{bmatrix} 0 & 1 & 1 & 1 \\ 1 & 0 & 1 & 1 \\ 1 & 1 & 0 & 1 \end{bmatrix}.$$

Encode n-vectors $\mathbf{x}$ by the $n+m$-vectors $\begin{bmatrix} \mathbf{x} \\ A\mathbf{x} \end{bmatrix}$. If an $n+m$-vector $\begin{bmatrix} \mathbf{x} \\ \mathbf{y} \end{bmatrix}$ is received, compute $A\mathbf{x} + \mathbf{y}$. If it is zero, the data vector was $\mathbf{x}$. If it is not zero, find which column of $[A|I]$ it is equal to. The index of this column indicates the erroneous bit.

The reader may conclude from the above discussion that "geometry" of $\mathbb{Z}_2$-vector spaces is strikingly different from that of $\mathbb{R}^2$ or $\mathbb{R}^3$ — $\mathbb{Z}_2$-spaces consist of only finitely many "points". In fact our geometrical arguments with affine spaces, graphs and linear operators work in $\mathbb{Z}_2$-spaces not because the geometry in $\mathbb{Z}_2^n$ is the same as in $\mathbb{R}^n$ but because the facts about $\mathbb{R}^n$ known to us from Section 3.3 rely exclusively on the axioms (i) – (vii) of abstract vector spaces which equally hold true in $\mathbb{R}^n$ and $\mathbb{Z}_2^n$.

Exercises 3.7.2.

(a) Our friend Professor Foulier from the College of Letters and Digits receives the message 1001001 encoded by the error-correcting code with the 3×4 matrix A above. What was the 4-bit string of data sent to him? Encode your answer by the 7-bit error-correcting code and check that it differs from the message received by Professor Foulier only in one bit.

(b) How many non-zero vectors, 1-dimensional subspaces, bases are there in $\mathbb{Z}_2^2$? How many invertible linear transformations from $\mathbb{Z}_2^2$ to itself are there?

3.7.3. Linear operators and ODEs. *Definition.* A linear operator from a vector space V to a vector space W is a function A from V to W which for any vectors $\mathbf{u}, \mathbf{v}$ from V and any scalars λ, μ satisfies the condition

$$A(\lambda\mathbf{u} + \mu\mathbf{v}) = \lambda A\mathbf{u} + \mu A\mathbf{v}.$$

Examples. (a) Suppose that V is finite-dimensional and therefore has a basis $\mathbf{f}_1, ..., \mathbf{f}_n$. Then any vector from V is uniquely written as a linear combination $\mathbf{v} = x_1\mathbf{f}_1 + ... + \mathbf{x}_n\mathbf{f}_n$. Let $\mathbf{w}_1, ..., \mathbf{w}_n$ be any n vectors from W. We define a linear operator A from V to W by putting $A\mathbf{v} = x_1\mathbf{w}_1 + ... + x_n\mathbf{w}_n$. Vice versa, any linear operator A from V to W has this form: $A\mathbf{v} = x_1A\mathbf{f}_1 + ... + x_nA\mathbf{f}_n$ is determined unambiguously by the n vectors $\mathbf{w}_1 = A\mathbf{f}_1, ..., \mathbf{w}_n = A\mathbf{f}_n$ in W. Suppose that W is also finite-dimensional. Then any vector from W is uniquely written as a linear combination $\mathbf{w} = y_1\mathbf{f}_1' + ... + y_m\mathbf{f}_m'$ of a basis $\mathbf{f}_1', ..., \mathbf{f}_m'$. Applying this to the vectors $\mathbf{w}_j = A\mathbf{f}_j$, $\mathbf{w}_j = a_{1j}\mathbf{f}_1' + ... + a_{mj}\mathbf{f}_m'$, $j = 1, ..., n$, we introduce the $m \times n$-matrix $A = [a_{ij}]$ and find that the linear operator A is described in coordinates $(x_1, ..., x_n)$ on V and $(y_1, ..., y_m)$ on W as the matrix product $\mathbf{y} = A\mathbf{x}$: $y_i = \sum_j a_{ij}x_j$.

(b) Linear operators from a $\mathbb{K}$-vector space V to $W = \mathbb{K}$ are called linear forms on V. Let V be the space of *continuous* functions on a closed interval $[a, b]$ (this is a subspace in the space $\mathbb{R}[a,b]$ of *all* such functions). Integration of continuous functions over $[a,b]$,

$$f \mapsto \int_a^b f(x)\, dx,$$

is a linear form on V since $\int(\lambda f(x)+\mu g(x))\, dx = \lambda \int f(x)\, dx + \mu \int g(x)\, dx$. More generally, one defines a linear form I_ϕ on V by picking a continuous function ϕ and putting

$$I_\phi f = \int_a^b \phi(x) f(x)\, dx.$$

Yet many linear forms on V can not be represented in this way. For instance, pick a point x_0 on $[a,b]$. Evaluation $f \mapsto f(x_0)$ of functions at x_0 is a linear form on V since $(\lambda f + \mu g)(x_0) = \lambda f(x_0) + \mu g(x_0)$. This suggests the formal notation

$$f(x_0) = \int_a^b \delta_{x_0}(x) f(x)\, dx,$$

where $\delta_{x_0}(x)$ is the Dirac δ-"function" equal 0 everywhere except x_0 and equal to infinity at x_0 (and describing the unit mass distribution of an ink spot concentrated at one point x_0).

(c) Let V be the space of infinitely differentiable functions on $\mathbb{R}$. Differentiation d/dt is a linear operator $Df = df/dt$ from V to itself since $(\lambda f(t)+\mu g(t))' = \lambda f'(t)+\mu g'(t)$. Linear operators from a vector space to itself are called **linear transformations** in the space. Given a linear transformation A, one can compose it with itself and introduce powers A^k and — more generally — polynomial expressions $a_0 A^n + a_1 A^{n-1} + ... + a_{n-1} A^1 + a_n A^0$. Let us consider the linear transformation $D^n + a_1 D^{n-1} + ... + a_n D^0$ in V. The null-space of this linear transformation is, by definition, the solution set to the ordinary differential equation

$$\frac{d^n}{dt^n} f + a_1 \frac{d^{n-1}}{dt^{n-1}} f + ... + a_n f = 0.$$

In particular, the solution set is a linear subspace in V. According to the Existence and Uniqueness Theorem the solution space has finite dimension n. As we found in Section 3.6.3, the functions

$$t^{k-1} e^{\lambda_i t},\ k = 1, ..., m_i,\ i = 1, ..., r, \text{ and}$$

$$t^{k-1} e^{a_j t} \cos b_j t,\ t^{k-1} e^{a_j t} \sin b_j t,\ k = 1, ..., l_j,\ j = 1, ..., s,$$

form a basis in this space provided that $\lambda_1, ..., \lambda_r, a_1 \pm i b_1, ..., a_s \pm i b_s$ are the complex roots of the polynomial $\lambda^n + a_1 \lambda^{n-1} + ... + a_n$, and $m_1, ..., m_r, l_1, ..., l_s$ are the corresponding multiplicities.

(d) Considering a more general linear operators defined by the formula $f \mapsto f^{(n)} + a_1(t) f^{(n-1)} + ... + a_n(t) f$ with non-constant (say, continuous) coefficient functions $a_1(t), ..., a_n(t)$, we conclude that the solution set of the corresponding linear homogeneous ODE $f^{(n)} + a_1(t) f + ... + a_n(t) f = 0$ is the null-space of this operator and has dimension n. Moreover, the solution set to an inhomogeneous linear ODE

$$\frac{d^n}{dt^n} f + a_1(t) \frac{d^{n-1}}{dt^{n-1}} f + ... + a_n(t) f = g(t)$$

is an affine subspace parallel to the null-space. In particular, the general solution to the differential equation has the form

$$f(t) = f_0(t) + c_1 f_1(t) + ... + c_n f_n(t),$$

where $f_1, ..., f_n$ form a basis of solutions to the corresponding homogeneous ODE, f_0 is a particular solution to the inhomogeneous ODE, and $c_1, ..., c_n$ are arbitrary constants. However, it is seldom possible to point out the functions $f_0(t), f_1(t), ..., f_n(t)$ explicitly (see Exercises though).

Exercises 3.7.3.

(a) Find all eigenvalues and eigenvectors of the linear transformation $D = \frac{d}{dt}$ in the space of infinitely differentiable functions on the number line.

(b)Suppose that the RHS of the differential equation

$$(D^n + a_1 D^{n-1} ... + a_n D^0) f = g$$

with constant coefficients $a_1, ..., a_n$ is a quasipolynomial of degree $\leq k$ of weight λ_0, that is a function of the form $e^{\lambda_0 t}(\alpha_0 + \alpha_1 t + ... + \alpha_k t^k)$. Prove that if

$$\lambda_0^n + a_1 \lambda_0^{n-1} + ... + a_n \neq 0,$$

then there exists a particular solution f_0 which is a quasipolynomial of the same weight and degree as g.

More generally, prove that if λ_0 is a multilplicity m root of the characteristic polynomial, then there exixts a particular solution f_0 which is a quasipolynomial of weight λ_0 and degree $\leq m + k$.

(c) Using (b) solve the ODEs:

$$f''' - 8f = t(e^t + e^{2t}),$$
$$f''' - 8f = \sin(\sqrt{3}t),$$
$$f''' - 8t = e^{-t} \sin(\sqrt{3}t).$$

3.7.4. The heat equation revisited. *Definition.* An inner product in a real vector space V is a function $\langle \mathbf{u}, \mathbf{v} \rangle$ of two vectors which satisfies symmetricity, bilinearity and positivity conditions: for any vectors $\mathbf{u}, \mathbf{v}, \mathbf{w}$ and any scalars λ, μ

$$\langle \mathbf{v}, \mathbf{u} \rangle = \langle \mathbf{u}, \mathbf{v} \rangle$$
$$\langle \lambda \mathbf{u} + \mu \mathbf{v}, \mathbf{w} \rangle = \lambda \langle \mathbf{u}, \mathbf{w} \rangle + \mu \langle \mathbf{v}, \mathbf{w} \rangle$$
$$\langle \mathbf{v}, \mathbf{v} \rangle > 0 \text{ unless } \mathbf{v} = \mathbf{0}.$$

A vector space provided with an inner product is called Euclidean. [10]

Examples. (a) Any finite-dimensional Euclidean vector space has an orthonormal basis. This follows from the Inertia Theorem applied to the positive-definite quadratic form $\langle \mathbf{x}, \mathbf{x} \rangle$.

(b) Let V be the space of infinitely differentiable 2π-periodic functions in one variable. The formula

$$\langle f, g \rangle = \int_{-\pi}^{\pi} f(x) g(x) \, dx$$

defines an inner product in V.

Definition. A linear transformation A in a Euclidean vector space is called symmetric if $\langle A\mathbf{u}, \mathbf{v} \rangle = \langle \mathbf{u}, A\mathbf{v} \rangle$ for any vectors $\mathbf{u}, \mathbf{v}$.

[10] In the case of complex vector spaces the first property of inner products should be modified: $\langle \mathbf{v}, \mathbf{u} \rangle = \overline{\langle \mathbf{u}, \mathbf{v} \rangle}$.

Examples. (c) If $\mathbf{f}_1, ..., \mathbf{f}_n$ is an orthonormal basis, then the matrix $[a_{ij}]$ of a symmetric linear transformation in this basis is symmetric, $a_{ij} = \langle \mathbf{f}_i, A\mathbf{f}_j\rangle = \langle A\mathbf{f}_i, \mathbf{f}_j\rangle = \langle \mathbf{f}_j, A\mathbf{f}_i\rangle = a_{ji}$, and vice versa.

(d) According to The Orthogonal Diagonalization Theorem, a symmetric linear transformation in a finite-dimensional Euclidean vector space has an orthogonal basis $\mathbf{v}_1, ..., \mathbf{v}_n$ of real eigenvectors: $A\mathbf{v}_i = \lambda_i \mathbf{v}_i$. Consider the system $\dot{\mathbf{x}} = A\mathbf{x}$ of linear ODEs with symmetric A. The solution, corresponding to a particular initial condition $\mathbf{x}(0)$, is described by the general formula

$$\mathbf{x}(t) = c_1 e^{\lambda_1 t}\mathbf{v}_1 + ... + c_n e^{\lambda_n t}\mathbf{v}_n,$$

where the values of the coefficients are given by

$$c_i = \langle \mathbf{x}(0), \mathbf{v}_i\rangle / \langle \mathbf{v}_i, \mathbf{v}_i\rangle.$$

The last formula is due to orthogonality of the eigenvectors:

$$\langle \mathbf{x}(0), \mathbf{v}_i\rangle = \sum c_j \langle \mathbf{v}_j, \mathbf{v}_i\rangle = c_i \langle \mathbf{v}_i, \mathbf{v}_i\rangle.$$

Consider the linear transformation $D^2 = \frac{d^2}{dx^2}$ in the Euclidean space V of infinitely differentiable 2π-periodic functions. We claim that it is symmetric. Indeed, since f, g, f', g' take on the same values at $x = \pm\pi$, we find

$$\langle D^2 f, g\rangle = \int_{-\pi}^{\pi} f''(x)g(x)dx = -\int_{-\pi}^{\pi} f'(x)g'(x)dx = \int_{-\pi}^{\pi} f(x)g''(x)dx = \langle f, D^2 g\rangle.$$

Let us look for eigenvectors of D^2, that is for non-zero 2π-periodic functions f satisfying the ODE $f'' = \lambda f$. For $\lambda > 0$ the solutions are linear combinations of $e^{\pm\sqrt{\lambda}t}$ and are not periodic. For $\lambda = 0$ the solutions have the form $a + bt$ and are periodic only when $b = 0$. If $\lambda = -\omega^2 < 0$, the solutions are linear combinations of $\cos\omega t$ and $\sin\omega t$. They are 2π-periodic only when $\omega = 1, 2, 3, ...$ Thus, the eigenvalues λ of D^2 are $0, -1, -4, -9, ...$, and the eigenspaces have dimensions $1, 2, 2, 2, ...$ Eigenvectors corresponding to different eigenvalues are orthogonal (why?). Since $\langle \cos nx, \sin nx\rangle = 0$ too, we conclude that the functions

$$\frac{1}{2}, \quad \cos t, \ \sin t, \quad \cos 2t, \ \sin 2t, \quad \cos 3t, \ \sin 3t, \ ...$$

form a complete set of pairwise orthogonal eigenfunctions, that is eigenvectors of D^2 in V.

Consider now the linear "system of ODEs" $\dot{\mathbf{u}} = A\mathbf{u}$ with the infinite - dimensional phase space V and with $A = D^2$. Since $\mathbf{u}$ here is a 2π-periodic function of x, the equation is in fact the heat equation $u_t = u_{xx}$ describing the heat conduction in a circular "wire" of length 2π with thermal diffusivity coefficient equal to 1. Driven by the analogy with Example (d), we obtain the solution to the heat equation in the form of a Fourier series

$$u(t,x) = \frac{a_0}{2} + \sum_{n=1}^{\infty} e^{-n^2 t}(a_n \cos nx + b_n \sin nx),$$

where the initial Fourier coefficients $a_0, a_1, b_1, ...$ are to be found from the initial condition: $a_0 = \langle u_0, 1/2\rangle / \langle 1/2, 1/2\rangle$,

$$a_n = \frac{\langle u_0, \cos nx\rangle}{\langle \cos nx, \cos nx\rangle}, \quad b_n = \frac{\langle u_0, \sin nx\rangle}{\langle \sin nx, \sin nx\rangle}.$$

Since the square lengths of the eigenfunctions are

$$\int_{-\pi}^{\pi} \frac{1}{4} dx = \frac{\pi}{2}, \int_{-\pi}^{\pi} \cos^2 nx \, dx = \pi, \int_{-\pi}^{\pi} \sin^2 nx \, dx = \pi,$$

we finally find

$$a_n = \frac{1}{\pi} \int_{-\pi}^{\pi} u(0,x) \cos nx \, dx, \; b_n = \frac{1}{\pi} \int_{-\pi}^{\pi} u(0,x) \sin nx \, dx.$$

In fact generalization of the Orthogonal Diagonalization Theorem to infinite dimensions is not straightforward (as a result of this, the solution formula works only for $t \geq 0$). Further discussion of this subject would be more appropriate for a functional analysis course.

Exercises 3.7.4.

(a) Prove that $\langle \cos nx, \sin nx \rangle = 0$.

(b) Modify the basis $1, \cos x, \sin x, \cos 2x, \sin 2x, \ldots$ of eigenfunctions of the transformation D^2 in order to make it ortho*normal.*

(c) Write down the (infinite!) matrix of the linear transformation $D = \frac{d}{dx}$ in the orthonormal basis of Exercise (a). Check that the matrix is anti-symmetric: $D^t = -D$.

(d) Show that the transformation $D = \frac{d}{dx}$ in the space V of infinitely differentiable 2π-periodic functions is anti-symmetric: $\langle Df, g \rangle = -\langle f, Dg \rangle$ for any f, g from V.

SAMPLE FINAL EXAM

1. Compute the rank of the linear operator from $\mathbb{R}^4$ to $\mathbb{R}^3$ given by the matrix

$$\begin{bmatrix} 2 & 1 & 3 & 7 \\ 3 & -2 & 1 & 7 \\ 1 & 4 & 5 & 7 \end{bmatrix}$$

and find a basis in the orthogonal complement to the null-space of this matrix.

2. Find the solution to the differential equation $f'''' + 2f'' + f = 0$ satisfying the initial condition $f(0) = f'(0) = f''(0) = 0,\ f'''(0) = 1$.

3. Find the temperature distribution $u(t, x)$ in a length-π solid bar with thermal diffusivity $\alpha^2 = 1$ if the initial temperature $u(0, x) = 1$ and the zero temperatures are maintained at the ends of the bar for $t > 0$.

4. Find the function $y = a + bx$ which provides the Least Square fitting to the experimental data

$$\begin{array}{cccccc} x & : & -1 & 0 & 1 & 2 \\ y & : & 2 & 1 & -1 & -3 \end{array}$$

5. Find out which of the following three matrices are similar to each other and which are not. (The problem requires some work. So, guess is not accepted - explain your decisions.)

$$A = \begin{bmatrix} 3 & -1 & -1 \\ -1 & 3 & -1 \\ -1 & -1 & 3 \end{bmatrix},\ B = \begin{bmatrix} 2 & 0 & \sqrt{2} \\ 0 & 4 & 0 \\ \sqrt{2} & 0 & 3 \end{bmatrix},\ C = \begin{bmatrix} 4 & 0 & 1 \\ 0 & 1 & 0 \\ 0 & 0 & 4 \end{bmatrix}.$$

6. Find out which of the following quadratic forms in $\mathbb{R}^3$ can be transformed into one another by linear changes of variables, and which cannot? Why?

$$P = 2xy + z^2,\quad Q = x^2 + y^2 + z^2 - 2xy - 2yz - 2zx,\quad R = 2xy + 2yz + 2zx$$

7. Formulate the definition of an eigenvector and prove that eigenvectors of a symmetric matrix corresponding to different eigenvalues are orthogonal to each other.

8. Is there a 3×3-matrix A such that $A^4 = \mathbf{0}$ but $A^3 \neq \mathbf{0}$? If "yes"— give an example, if "no" — explain "why".